A strophysics in a Nutshell

Astrophysics in a Nutshell

Dan Maoz

PRINCETON UNIVERSITY PRESS · PRINCETON AND OXFORD

Copyright © 2007 by Princeton University Press

Published by Princeton University Press, 41 William Street, Princeton, New Jersey 08540

In the United Kingdom: Princeton University Press, 3 Market Place, Woodstock, Oxfordshire OX20 1SY

Library of Congress Cataloging-in-Publication Data

Maoz, Dan.
    Astrophysics in a nutshell / Dan Maoz.
      p. cm.
    Includes bibliographical references and index.
    ISBN-13: 978-0-691-12584-8 (acid-free paper)
    ISBN-10: 0-691-12584-8 (acid-free paper)
      1. Astrophysics.   I. Title.
  QB461.M32 2007
  523.01—dc22   2006050588

British Library Cataloging-in-Publication Data is available

This book has been composed in Scala LF and Scala Sans

Printed on acid-free paper. ∞

press.princeton.edu

Printed in the United States of America

10 9 8 7 6 5 4

To Orit, Lia and Yonatan—
the three bright stars in my sky; and to my parents

Contents

Preface

This textbook is based on the one-semester course "Introduction to Astrophysics," taken by third-year physics students at Tel-Aviv University, which I taught several times in the years 2000–2005. My objective in writing this book is to provide an introductory astronomy text that is suited for university students majoring in physical science fields (physics, astronomy, chemistry, engineering, etc.), rather than for a wider audience, for which many astronomy textbooks already exist. I have tried to cover a large and representative fraction of the main elements of modern astrophysics, including some topics at the forefront of current research. At the same time, I have made an effort to keep this book concise.

I covered this material in approximately forty 45-minute lectures. The text assumes a level of math and physics expected from intermediate-to-advanced undergraduate science majors, namely, familiarity with calculus and differential equations, classical and quantum mechanics, special relativity, waves, statistical mechanics, and thermodynamics. However, I have made an effort to avoid long mathematical derivations, or complicated physical arguments that might mask simple realities. Thus, throughout the text, I use devices such as scaling arguments and order-of-magnitude estimates to arrive at the important basic results. Where relevant, I then state the results of more thorough calculations that involve, e.g., taking into account secondary processes that I have ignored, or full solutions of integrals, or of differential equations.

Undergraduates are often taken aback by their first encounter with this order-of-magnitude approach. Of course, full and accurate calculations are as indispensable in astrophysics as in any other branch of physics (e.g., an omitted factor of π may not be important for understanding the underlying physics of some phenomenon, but it can be very important for comparing a theoretical calculation to the results of an experiment). However, most physicists (regardless of subdiscipline), when faced with a new problem, will first carry out a rough, "back-of-the-envelope" analysis that can lead to some basic intuition about the processes and the numbers involved. Thus, the approach we will follow here is actually valuable and widely used, and the student is well advised to attempt to become proficient at it. With this objective in mind, some derivations and some topics are left as problems at the end of each chapter (usually including a generous amount of guidance), and solving most or all of the problems is highly recommended in order to get the most out of this book. I have not provided full solutions to the problems, to counter the temptation to peek. Instead, at the end of some problems I have provided short answers that permit the reader to check the correctness of the solution, although not in cases where the answer would give away the solution too easily (physical science students are notoriously competent at "reverse engineering" a solution—not necessarily correct—to an answer!).

There is much that does *not* appear in this book. I have excluded almost all descriptions of the historical developments of the various topics, and have, in general, presented them directly as they are understood today. There is almost no attribution of results to the many scientists whose often-heroic work has led to this understanding, a choice that certainly does injustice to many individuals, past and living. Furthermore, not all topics in astrophysics are equally amenable to the type of exposition this book follows, and I naturally have my personal biases about what is most interesting and important. As a result, the coverage of the different subjects is intentionally uneven: some are explored to considerable depth, while others are presented only descriptively, given brief mention, or completely omitted. Similarly, in some cases I develop from "first principles" the physics required to describe a problem, but in other cases I begin by simply stating the physical result, either because I expect the reader is already familiar enough with it, or because developing it would take too long. I believe that all these choices are essential to keep the book concise, focused, and within the scope of a one-term course. No doubt, many people will disagree with the particular choices I have made, but hopefully will agree that all that has been omitted here can be covered later by more advanced courses (and the reader should be aware that proper attribution of results *is* the strict rule in the research literature).

Astronomers use some strange units, in some cases for no reason other than tradition. I generally use cgs units, but also make frequent use of some other units that are common in astronomy, e.g., angstroms, kilometers, parsecs, light-years, years, solar masses, and solar luminosities. However, I have completely avoided using or mentioning "magnitudes," the peculiar logarithmic units used by astronomers to quantify flux. Although magnitudes are widely used in the field, they are not required for explaining anything in this book, and might only cloud the real issues. Again, students continuing to more advanced courses and to research can easily deal with magnitudes at that stage.

A note on equality symbols and their relatives. I use an "=" sign, in addition to cases of strict mathematical equality, for numerical results that are accurate to better than ten percent. Indeed, throughout the text I use constants and unit transformations with only two significant digits (they are also listed in this form in "Constants and Units," in the hope that the student will memorize the most commonly used among them after a while), except in a few places where more digits are essential. An "$\approx$" sign in a mathematical relation (i.e., when mathematical symbols, rather than numbers, appear on both sides) means some approximation has been made, and in a numerical relation it means an accuracy somewhat worse than ten percent. A "$\propto$" sign means strict proportionality between the two sides. A "$\sim$" is used in two senses. In a mathematical relation it means an approximate functional dependence. For example, if $y = ax^{2.2}$, I may write $y \sim x^2$. In numerical relations, I use "$\sim$" to indicate order-of-magnitude accuracy.

This book has benefitted immeasurably from the input of the following colleagues, to whom I am grateful for providing content, comments, insights, ideas, and many corrections: T. Alexander, R. Barkana, M. Bartelmann, J.-P. Beaulieu, D. Bennett, D. Bram, D. Champion, M. Dominik, H. Falcke, A. Gal-Yam, A. Ghez, O. Gnat, A. Gould, B. Griswold, Y. Hoffman, S. Jha, M. Kamionkowski, S. Kaspi, V. Kaspi, A. Laor, A. Levinson, J. R. Lu, J. Maos, T. Mazeh, J. Peacock, D. Poznanski, P. Saha, D. Spergel, A. Sternberg, R. Thompson, R. Webbink, L. R. Williams, and S. Zucker. The remaining errors are, of course, all my own. Orit Bergman patiently produced most of the figures—one more of the many things she has granted me, and for which I am forever thankful.

D.M.

Tel-Aviv, 2006

Constants and Units (to two significant digits)

Gravitational constant	$G = 6.7 \times 10^{-8}$ erg cm g^{-2}
Speed of light	$c = 3.0 \times 10^{10}$ cm s^{-1}
Planck's constant	$h = 6.6 \times 10^{-27}$ erg s
	$\hbar = h/2\pi = 1.05 \times 10^{-27}$ erg s
Boltzmann's constant	$k = 1.4 \times 10^{-16}$ erg K^{-1}
	$= 8.6 \times 10^{-5}$ eV K^{-1}
Stefan-Boltzmann constant	$\sigma = 5.7 \times 10^{-5}$ erg cm^{-2} s^{-1} K^{-4}
Radiation constant	$a = 4\sigma/c = 7.6 \times 10^{-15}$ erg cm^{-3}K^{-4}
Proton mass	$m_p = 1.7 \times 10^{-24}$ g
Electron mass	$m_e = 9.1 \times 10^{-28}$ g
Electron charge	$e = 4.8 \times 10^{-10}$ esu
Electron volt	1 eV $= 1.6 \times 10^{-12}$ erg
Thomson cross section	$\sigma_T = 6.7 \times 10^{-25}$ cm^2
Wien's law	$\lambda_{max} = 2900$ Å 10^4 K$/T$
	$h\nu_{max} = 2.4$ eV $T/10^4$ K
Angstrom	1 Å $= 10^{-8}$ cm
Solar mass	$M_\odot = 2.0 \times 10^{33}$ g
Solar luminosity	$L_\odot = 3.8 \times 10^{33}$ erg s^{-1}
Solar radius	$r_\odot = 7.0 \times 10^{10}$ cm

Solar distance $\qquad\qquad d_\odot = 1\text{ AU} = 1.5 \times 10^{13}\text{ cm}$

Jupiter mass $\qquad\qquad M_J = 1.9 \times 10^{30}\text{ g}$

Jupiter radius $\qquad\qquad r_J = 7.1 \times 10^9\text{ cm}$

Jupiter distance $\qquad\qquad d_J = 5\text{ AU} = 7.5 \times 10^{13}\text{ cm}$

Earth mass $\qquad\qquad M_\oplus = 6.0 \times 10^{27}\text{ g}$

Earth radius $\qquad\qquad r_\oplus = 6.4 \times 10^8\text{ cm}$

Moon mass $\qquad\qquad M_{\text{moon}} = 7.4 \times 10^{25}\text{ g}$

Moon radius $\qquad\qquad r_{\text{moon}} = 1.7 \times 10^8\text{ cm}$

Moon distance $\qquad\qquad d_{\text{moon}} = 3.8 \times 10^{10}\text{ cm}$

Astronomical unit $\qquad\qquad 1\text{ AU} = 1.5 \times 10^{13}\text{ cm}$

Parsec $\qquad\qquad 1\text{ pc} = 3.1 \times 10^{18}\text{ cm} = 3.3\text{ ly}$

Year $\qquad\qquad 1\text{ yr} = 3.15 \times 10^7\text{ s}$

A strophysics in a Nutshell

1 | Introduction

Astrophysics is the branch of physics that studies, loosely speaking, phenomena on large scales—the Sun, the planets, stars, galaxies, and the Universe as a whole. But this definition is clearly incomplete; much of astronomy[1] also deals, e.g., with phenomena at the atomic and nuclear levels. We could attempt to define astrophysics as the physics of distant objects and phenomena, but astrophysics also includes the formation of the Earth, and the effects of astronomical events on the emergence and evolution of life on Earth. This semantic difficulty perhaps simply reflects the huge variety of physical phenomena encompassed by astrophysics. Indeed, as we will see, practically all the subjects encountered in a standard undergraduate physical science curriculum—classical mechanics, electromagnetism, thermodynamics, quantum mechanics, statistical mechanics, relativity, and chemistry, to name just some—play a prominent role in astronomical phenomena. Seeing all of them in action is one of the exciting aspects of studying astrophyics.

Like other branches of physics, astronomy involves an interplay between experiment and theory. Theoretical astrophysics is carried out with the same tools and approaches used by other theoretical branches of physics. Experimental astrophysics, however, is somewhat different from other experimental disciplines, in the sense that astronomers cannot carry out controlled experiments,[2] but can only perform **observations** of the various phenomena provided by nature. With this in mind, there is little difference, in practice, between the design and the execution of an experiment in some field of physics, on the one hand, and the design and the execution of an astronomical observation, on the other. There is certainly no particular distinction between the methods of data analysis in either case. But, since everything we discuss in this book will ultimately be based on observations, let us begin with a brief overview of how observations are used to make astrophysical measurements.

---

[1] We will use the words "astrophysics" and "astronomy" interchangeably, as they mean the same thing nowadays. For example, the four leading journals in which astrophysics research is published are named *The Astrophysical Journal*, *The Astronomical Journal*, *Astronomy and Astrophysics*, and *Monthly Notices of the Royal Astronomical Society*, but their subject content is the same.

[2] An exception is the field of laboratory astrophysics, in which some particular properties of astronomical conditions are simulated in the lab.

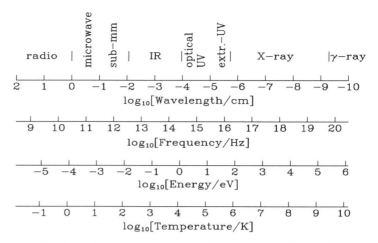

Figure 1.1 The various spectral regions of electromagnetic radiation, their common astronomical nomenclature, and their approximate borders in terms of wavelength, frequency, energy, and temperature. Temperature is here associated with photon energy E via the relation $E = kT$, where k is Boltzmann's constant.

1.1 Observational Techniques

With several exceptions, astronomical phenomena are almost always observed by detecting and measuring electromagnetic (EM) radiation from distant sources. (The exceptions are in the fields of cosmic ray astronomy, neutrino astronomy, and gravitational wave astronomy.) Figure 1.1 shows the various, roughly defined, regions of the EM spectrum. To record and characterize EM radiation, one needs, at least, a camera that will focus the approximately plane EM waves arriving from a distant source and a detector at the focal plane of the camera, which will record the signal. A "telescope" is just another name for a camera that is specialized for viewing distant objects. The most basic such camera–detector combination is the human eye, which consists (among other things) of a lens (the camera) that focuses images on the retina (the detector). Light-sensitive cells on the retina then translate the light intensity of the images into nerve signals that are transmitted to the brain. Figure 1.2 sketches the optical principles of the eye and of two telescope configurations.

Until the introduction of telescope use to astronomy by Galileo in 1609, observational astronomy was carried out solely using human eyes. However, the eye as an astronomical tool has several disadvantages. The **aperture** of a dark-adapted pupil is <1 cm in diameter, providing limited **light-gathering area** and limited **angular resolution**. The light-gathering capability of a camera is set by the area of its aperture (e.g., of the objective lens, or of the primary mirror in a reflecting telescope). The larger the aperture, the more photons, per unit time, can be detected, and hence fainter sources of light can be observed. For example, the largest visible-light telescopes in operation today have 10-meter primary mirrors, i.e., more than a million times the light gathering area of a human eye.

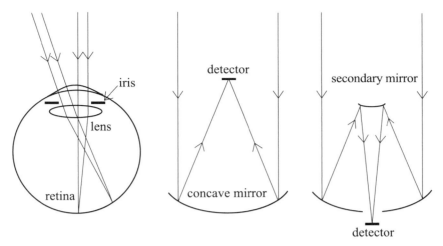

Figure 1.2 Optical sketches of three different examples of camera-detector combinations. *Left:* Human eye, shown with parallel rays from two distant sources, one source on the optical axis of the lens and one at an angle to the optical axis. The lens, which serves as the camera in this case, focuses the light onto the retina (the detector), on which two point images are formed. *Center:* A reflecting telecope with a detector at its *prime focus*. Plotted are parallel rays from a distant source on the optical axis of the telescope. The concave mirror focuses the rays onto the detector at the mirror's focal plane, where a point image is formed. *Right:* Reflecting telescope, but with a secondary, convex, mirror, which folds the beam back down and through a hole in the primary concave mirror, to form an image on the detector at the so-called *Cassegrain focus*.

The angular resolution of a camera or a telescope is the smallest angle on the sky between two sources of light that can be discerned as separate sources with that camera. From wave optics, a plane wave of wavelength λ passing through a circular aperture of diameter D, when focused onto a detector, will produce a diffraction pattern of concentric rings, centered on the position expected from geometrical optics, with a central spot having an angular radius (in radians) of

$$\theta = 1.22 \frac{\lambda}{D}. \tag{1.1}$$

Consider, for example, the image of a field of stars obtained through some camera, and having also a bandpass filter that lets through light only within a narrow range of wavelengths. The image will consist of a set of such diffraction patterns, one at the position of each star (see Fig. 1.3). Actually seeing these diffraction patterns requires that blurring of the image not be introduced, either by imperfectly built optics or by other elements, e.g., Earth's atmosphere. The central spots from the diffraction patterns of two adjacent sources on the sky will overlap, and will therefore be hard to distinguish from each other, when their angular separation is less than about λ/D. Similarly, a source of light with an intrinsic angular size smaller than this **diffraction limit** will produce an image that is *unresolved*, i.e., indistinguishable from the image produced by a **point source** of zero angular extent. Thus, in principle, a 10-meter telescope working at the same visual wavelengths as the eye can have an angular resolution that is 1000 times better than that of the eye.

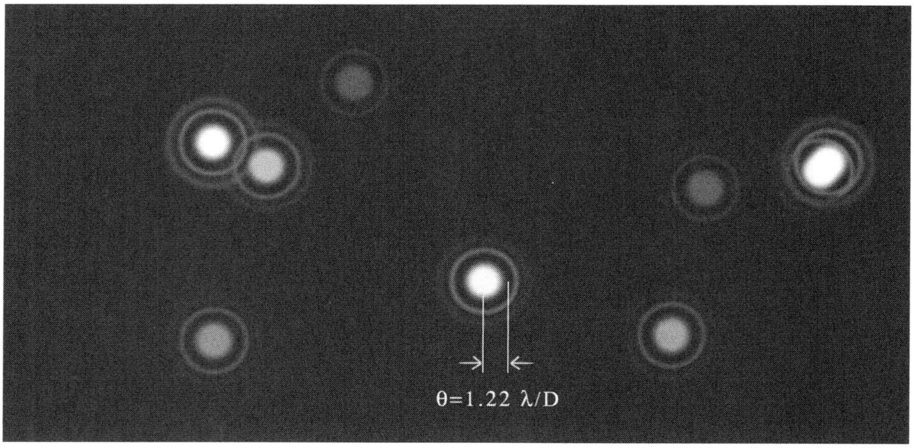

Figure 1.3 Simulated diffraction-limited image of a field of stars, with the characteristic diffraction pattern due to the telescope's finite circular aperture at the position of every star. Pairs of stars separated on the sky by an angle $\theta < \lambda/D$ (e.g., on the right-hand side of the image) are hard to distinguish from single stars. Real conditions are always worse than the diffraction limit, due to, e.g., imperfect optics and atmospheric blurring.

In practice, it is difficult to achieve diffraction-limited performance with ground-based *optical* telescopes, due to the constantly changing, blurring effect of the atmosphere. (The **optical** wavelength range of EM radiation is roughly defined as $0.32–1$ μm.) However, observations with angular resolutions at the diffraction limit are routine in radio and infrared astronomy, and much progress in this field has been achieved recently in the optical range as well. Angular resolution is important not only for discerning the fine details of astronomical sources (e.g., seeing the moons and surface features of Jupiter, the constituents of a star-forming region, or subtle details in a galaxy), but also for detecting faint unresolved sources against the background of emission from the Earth's atmosphere, i.e., the "sky." The night sky shines due to scattered light from the stars, from the Moon, if it is up, and from artificial light sources, but also due to fluorescence of atoms and molecules in the atmosphere. The better the angular resolution of a telescope, the smaller the solid angle over which the light from, say, a star, will be spread out, and hence the higher the contrast of that star's image over the statistical fluctuations of the sky background (see Fig. 1.4). A high sky background combined with limited angular resolution are among the reasons why it is difficult to see stars during daytime.

A third limitation of the human eye is its fixed integration time, of about 1/30 second. In astronomical observations, faint signals can be collected on a detector during arbitrarily long exposures (sometimes accumulating to months), permitting the detection of extremely faint sources. Another shortcoming of the human eye is that it is sensitive only to a narrow **visual** range of wavelengths of EM radiation (about $0.4–0.7$ μm, i.e., within the optical range defined above), while astronomical information exists in all regions of the EM spectrum, from radio, through infrared, optical, ultraviolet, X-ray, and gamma-ray bands. Finally, a detector other than the eye allows keeping an objective record of the observation,

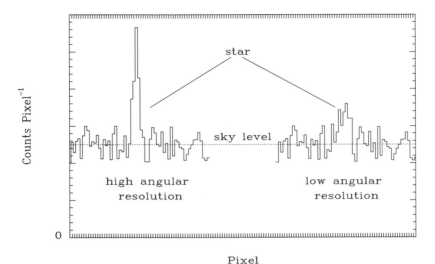

Figure 1.4 Cuts through the positions of a star in two different astronomical images, illustrating the effect of angular resolution on the detectability of faint sources on a high background. The vertical axis shows the counts registered in every pixel along the cut, as a result of the light intensity falling on that pixel. On the left, the narrow profile of the stellar image stands out clearly above the Poisson fluctuations in the sky background, the mean level of which is indicated by the dashed line. On the right, the counts from the same star are spread out in a profile that is twice as wide, and hence the contrast above the background noise is lower.

which can then be examined, analyzed, and disseminated among other researchers. Astronomical data are almost always saved in some digital format, in which they are most readily later processed using computers. All telescopes used nowadays for professional astronomy are equipped with detectors that record the data (whether an image of a section of sky, or otherwise—see below). The popular perception of astronomers peering through the eyepieces of large telescopes is a fiction.

The type of detector that is used in optical, near-ultraviolet, and X-ray astronomy is almost always a **charge-coupled device** (CCD), the same type of detector that is found in commercially available digital cameras. A CCD is a slab of silicon that is divided into numerous *pixels* by a combination of insulating buffers that are etched into the slab and the application of selected voltage differences along its area. Photons reaching the CCD liberate *photoelectrons* via the photoelectric effect. The photoelectrons accumulated in every pixel during an exposure period are then read out and amplified, and the measurement of the resulting current is proportional to the number of photons that reached the pixel. This allows forming a digital image of the region of the sky that was observed (see Fig. 1.5).

So far, we have discussed astronomical observations only in terms of producing an image of a section of sky by focusing it onto a detector. This technique is called *imaging*. However, an assortment of other measurements can be made. Every one of the parameters that characterize an EM wave can carry useful astronomical information. Different techniques

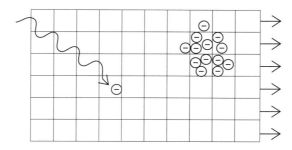

Figure 1.5 Schematic view (highly simplified) of a CCD detector. On the left, a photon is absorbed by the silicon in a particular pixel, releasing an electron, which is stored in the pixel until the CCD is read out. On the right are shown other photoelectrons that were previously liberated and stored in several pixels on which, e.g., the image of a star has been focused. At the end of the exposure, the accumulated charge is transferred horizontally from pixel to pixel by manipulating the voltages applied to the pixels, until it is read out on the right-hand side (arrows) and amplified.

have been designed to measure each of these parameters. To see how, consider a plane-parallel, monochromatic (i.e., having a single frequency), EM wave, with electric field vector described by

$$\mathbf{E} = \hat{\mathbf{e}} E(t) \cos{(2\pi \nu t - \mathbf{k} \cdot \mathbf{r} + \phi)}. \tag{1.2}$$

The unit vector $\hat{\mathbf{e}}$ gives the direction of polarization of the electric field, $E(t)$ is the field's time-dependent (apart from the sinusoidal variation) amplitude, ν is the frequency, and $\mathbf{k}$ is the wave vector, having the direction of the wave propagation, and magnitude $|\mathbf{k}| = 2\pi/\lambda$. The wavelength λ and the frequency ν are related by the speed of light, c, through $\nu = c/\lambda$. The phase shift of the wave is ϕ.

Imaging involves determining the direction, on the sky, to a source of plane-parallel waves, and therefore implies a measurement of the direction of $\mathbf{k}$. From an image, one can also measure the strength of the signal produced by a source (e.g., in a photon-counting device, by counting the total number of photons collected from the source over an integration time). As discussed in more detail in chapter 2, the photon flux is related to the *intensity*, which is the time-averaged electric-field amplitude squared, $\langle E^2(t) \rangle$. Measuring the photon flux from a source is called *photometry*. In time-resolved photometry, one can perform repeated photometric measurements as a function of time, and thus measure the long-term time dependence of $\langle E^2 \rangle$.

The wavelength of the light, λ (or equivalently, the frequency, ν), can be determined in several ways. A bandpass filter before the detector (or in the "receiver" in radio astronomy) will allow only EM radiation in a particular range of wavelengths to reach the detector, while blocking all others. Alternatively, the light can be reflected off, or transmitted through, a dispersing element, such as a prism or a diffraction grating, before reaching the detector. Light of different wavelengths will be deflected by different angles from the original beam, and hence will land on the detector at different positions. A single source

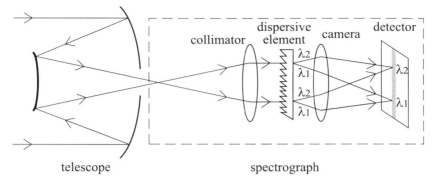

Figure 1.6 Schematic example of a spectrograph. Light from a distant point source converges at the Cassegrain focus of the telecope at the left. The beam is then allowed to diverge again and reaches a *collimator* lens sharing the same focus as the telecope, so that a parallel beam of light emerges. The beam is then transmitted through a dispersive element, e.g., a transmission grating, which deflects light of different wavelengths by different angles, in proportion to the wavelength. The paths of rays for two particular wavelengths, λ_1 and λ_2, are shown. A camera lens refocuses the light onto a detector at the camera's focal plane. The light from the source, rather than being imaged into a point, has been spread into a spectrum (gray vertical strip).

of light will thus be spread into a **spectrum**, with the signal at each position along the spectrum proportional to the intensity at a different wavelength. This technique is called *spectroscopy,* and an example of a telescope–spectrograph combination is illustrated in Fig. 1.6.

The phase shift ϕ of the light wave arriving at the detector can reveal information on the precise direction to the source and on effects, such as scattering, that the wave underwent during its path from the source to the detector. The phase can be measured by combining the EM waves received from the same source by several different telescopes and forming an interference pattern. This is called *interferometry.* In interferometry, the *baseline* distance B between the two most widely spaced telescopes replaces the aperture in determining the angular resolution, λ/B. In radio astronomy, the signals from radio telescopes spread over the globe, and even in space, are often combined, providing baselines of order 10^4 km, and very high angular resolutions.

Finally, the amount of polarization (*unpolarized,* i.e., having random polarization direction, or *polarized* by a fraction between 0 and 100%), its type (linear, circular), and the orientation on the sky of the polarization vector $\hat{\mathbf{e}}$ can be determined. For example, in optical astronomy this can be achieved by placing polarizing filters in the light beam, allowing only a particular polarization component to reach the detector. Measurement of the polarization properties of a source is called *polarimetry.*

Ideally, one would like always to be able to characterize all of the parameters of the EM waves from a source, but this is rarely feasible in practice. Nevertheless, it is often possible to measure several characteristics simultaneously, and these techniques are then referred to by the appropriate names, e.g., spectro-photo-polarimetry, in which both the intensity and the polarization of light from a source are measured as a function of wavelength.

In the coming chapters, we study some of the main topics with which astrophysics deals, generally progressing from the near to the far. Most of the volume of this book is dedicated to the theoretical understanding of astronomical phenomena. However, it is important to remember that the discovery and quantification of those phenomena are the products of observations, using the techniques that we have just briefly reviewed.

Problems

1. a. Calculate the best angular resolution that can, in principle, be achieved with the human eye. Assume a pupil diameter of 0.5 cm and the wavelength of green light, ~0.5 μm. Express your answer in arcminutes, where an arcminute is 1/60 of a degree. (In practice, the human eye does not achieve diffraction-limited performance, because of imperfections in the eye's optics and the coarse sampling of the retina by the light-sensitive *rod* and *cone* cells that line it.)

   b. What is the angular resolution, in arcseconds (1/3600 of a degree), of the Hubble Space Telescope (with an aperture diameter of 2.4 m) at a wavelength of 0.5 μm?

   c. What is the angular resolution, expressed as a fraction of an arcsecond, of the Very Long Baseline Interferometer (VLBI)? VLBI is an network of radio telescopes (wavelengths ~1–100 cm), spread over the globe, that combine their signals to form one large interferometer.

   d. From the table of Constants and Units, find the distances and physical sizes of the Sun, Jupiter, and a Sun-like star 10 light years away. Calculate their angular sizes, and compare to the angular resolutions you found above.

2. A CCD detector at the focal plane of a 1-m-diameter telescope records the image of a certain star. Due to the blurring effect of the atmosphere (this is called "seeing" by astronomers) the light from the star is spread over a circular area of radius R pixels. The total number of photoelectrons over this area, accumulated during the exposure, and due to the light of the star, is N_{star}. Light from the sky produces n_{sky} photoelectrons per pixel in the same exposure.

   a. Calculate the signal-to-noise ratio (S/N) of the photometric measurement of the star, i.e., the ratio of the counts from the star to the uncertainty in this measurement. Assume Poisson statistics, i.e., that the "noise" is the square root of the total counts, from all sources.

   b. The same star is observed with the same exposure time, but with a 10-m-diameter telescope. This larger telescope naturally has a larger light gathering area, but also is at a site with a more stable atmosphere, and therefore has 3 times better "seeing" (i.e., the light from the stars is spread over an area of radius $R/3$). Find the S/N in this case.

   c. Assuming that the star and the sky are not variable (i.e., photons arrive from them at a constant rate), find the functional dependence of S/N on exposure time, t, in two limiting cases: the counts from the star are much greater than the counts from the sky in the "seeing disk"; and vice versa.

Answer: $S/N \propto t^{1/2}$ in both cases.

d. Based on the results of (c), by what factor does the exposure time with the 1-m telescope need to be increased to reach the S/N obtained with the 10-m telescope, for each of the two limiting cases?

Answer: By a factor 100 in the first case, and 1000 in the second case.

2 | Stars: Basic Observations

In this chapter we examine some of the basic observed properties of stars—their spectra, temperatures, emitted power, and masses—and the relations between those properties. In chapter 3, we proceed to a physical understanding of these observations.

2.1 Review of Blackbody Radiation

To a very rough, but quite useful, approximation, stars shine with the spectrum of a **blackbody**. The degree of similarity (but also the differences) between stellar and blackbody spectra can be seen in Fig. 2.1. Let us review the various descriptions and properties of blackbody radiation (which is often also called *thermal radiation*, or radiation having a *Planck spectrum*). A blackbody spectrum emerges from a system in which matter and radiation are in thermodynamic equilibrium. A fundamental result of quantum mechanics (and one that marked the beginning of the quantum era in 1900) is the exact functional form of this spectrum, which can be expressed in a number of ways.

The **energy density** of blackbody radiation, per frequency interval, is

$$u_\nu = \frac{8\pi \nu^2}{c^3} \frac{h\nu}{e^{h\nu/kT} - 1}, \tag{2.1}$$

where ν is the frequency, c is the speed of light, h is Planck's constant, k is Boltzmann's constant, and T is the temperature in degrees Kelvin. Clearly, the first term has units of [time]/[length]^3 and the second term has units of energy. In cgs units, u_ν is given in erg cm^{-3} Hz^{-1}.

Next, let us consider the flow of blackbody energy radiation (i.e., photons moving at speed c), in a particular direction inside a blackbody radiator. To obtain this so-called **intensity**, we take the derivative with respect to solid angle of the energy density and

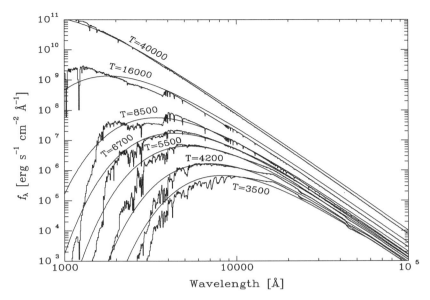

Figure 2.1 Flux per wavelength interval emitted by different types of stars, at their "surfaces," compared to blackbody curves of various temperatures. Each blackbody's temperature is chosen to match the total power (integrated over all wavelengths) under the the corresponding stellar spectrum. The wavelength range shown is from the ultraviolet (1000 Å = 0.1 μm), through the optical range (3200–10,000 Å), and to the mid-infrared (10^5 Å = 10 μm). Data credit: R. Kurucz.

multiply by c (since multiplying a density by a velocity gives a flux, i.e., the amount passing through a unit area per unit time):

$$I_\nu = c \frac{du_\nu}{d\Omega}, \tag{2.2}$$

where $d\Omega$ is the solid angle element. (For example, in spherical coordinates, $d\Omega = \sin\theta\, d\theta\, d\phi$.) Blackbody radiation is isotropic (i.e., the same in all directions), and hence the energy density per unit solid angle is

$$\frac{du_\nu}{d\Omega} = \frac{u_\nu}{4\pi} \tag{2.3}$$

(since the solid angle of a full sphere is 4π steradians). The intensity of blackbody radiation is therefore

$$I_\nu = \frac{c}{4\pi} u_\nu = \frac{2h\nu^3}{c^2} \frac{1}{e^{h\nu/kT} - 1} \equiv B_\nu. \tag{2.4}$$

In cgs, one can see the units now are erg s^{-1} cm^{-2} Hz^{-1} steradian^{-1}. We have kept the product of units, s^{-1} Hz^{-1}, even though they formally cancel out, to recall their different physical origins: one is the time interval over which we are measuring the amount of energy that flows through a unit area; and the other is the photon frequency interval over which we bin the spectral distribution. I_ν of a blackbody is often designated B_ν.

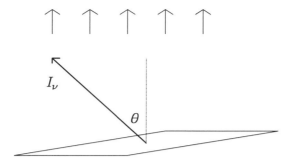

Figure 2.2 Illustration of the net flux emerging through surface of a blackbody, due to a beam with intensity I_ν emerging at an angle θ to the perpendicular.

Now, let us find the net flow of energy that emerges from a unit area (small enough so that it can be presumed to be flat) on the outer surface of a blackbody (see Fig. 2.2). This is obtained by integrating I_ν over solid angle on the half-sphere facing outward, with each I_ν weighted by the cosine of the angle between the intensity and the perpendicular to the area. This **flux**, which is generally what one actually observes from stars and other astronomical sources, is thus

$$f_\nu = \int_{\theta=0}^{\pi/2} I_\nu \cos\theta \, d\Omega = I_\nu 2\pi \frac{1}{2} = \pi I_\nu = \frac{c}{4} u_\nu = \frac{2\pi h\nu^3}{c^2} \frac{1}{e^{h\nu/kT} - 1}. \tag{2.5}$$

The cgs units of this flux per frequency interval will thus be erg s^{-1} cm^{-2} Hz^{-1}.

The total power (i.e., the energy per unit time) radiated by a spherical, isotropically emitting, star of radius r_* is usually called its **luminosity**, and is just

$$L_\nu = f_\nu(r_*) 4\pi r_*^2, \tag{2.6}$$

with cgs units of erg s^{-1} Hz^{-1}. Similarly, the flux that an observer at a distance d from the star will measure will be

$$f_\nu(d) = \frac{L_\nu}{4\pi d^2} = f_\nu(r_*) \frac{r_*^2}{d^2}. \tag{2.7}$$

It is often of interest to consider the above quantities integrated over all photon frequencies, and designated by

$$u = \int_0^\infty u_\nu \, d\nu, \quad I = \int_0^\infty I_\nu \, d\nu, \quad f = \int_0^\infty f_\nu \, d\nu, \quad L = \int_0^\infty L_\nu \, d\nu. \tag{2.8}$$

A case in point is the useful **Stefan-Boltzmann law**, which relates the total energy density or flux of a blackbody to its temperature:

$$u = aT^4, \tag{2.9}$$

and

$$f = \frac{c}{4} a T^4 = \sigma T^4, \tag{2.10}$$

$$a = \frac{8\pi^5 k^4}{15 c^3 h^3} = 7.6 \times 10^{-15} \text{ erg cm}^{-3} \text{K}^{-4},$$

$$\sigma = \frac{c}{4} a = 5.7 \times 10^{-5} \text{ erg s}^{-1} \text{ cm}^{-2} \text{K}^{-4}.$$

(Here and throughout this book, numbers are rounded off to two significant digits, except in some obvious cases where higher accuracies are warranted.)

Rather than considering energy density, intensity, flux, and luminosity per photon frequency interval, we can also look at these quantities per photon wavelength interval, where the wavelength is $\lambda = c/\nu$. To make the transformation, we recall that the energy in an interval must be the same, whether we measure it in wavelength or frequency, so,

$$B_\lambda d\lambda = B_\nu d\nu, \tag{2.11}$$

and hence

$$B_\lambda = B_\nu \left| \frac{d\nu}{d\lambda} \right| = B_\nu \frac{c}{\lambda^2} = \frac{2hc^2}{\lambda^5} \frac{1}{e^{hc/\lambda kT} - 1}. \tag{2.12}$$

Here the units are erg s^{-1} cm^{-2} cm^{-1} steradian^{-1}, where we have separated the two length units (cm^{-2} and cm^{-1}), since one is the unit area through which the radiation flux is passing, and the other is the wavelength interval over which we bin the radiation energy. Non-cgs units for the wavelength interval are common in astronomy. For example, flux per wavelength interval at visual wavelengths is often given in units of erg s^{-1} cm^{-2} Å^{-1}. An Å (called "angstrom") is 10^{-8} cm.

The wavelength or frequency of the peak of a blackbody spectrum can be found by taking its derivative and equating to zero:

$$\frac{dB_\nu}{d\nu} = 0, \tag{2.13}$$

or

$$\frac{dB_\lambda}{d\lambda} = 0, \tag{2.14}$$

which lead to the two forms of **Wien's law**:

$$\lambda_{\max} T = 0.29 \text{ cm K} \tag{2.15}$$

and

$$h\nu_{\max} = 2.8\, kT. \tag{2.16}$$

For example, the nearest star—the Sun—which radiates approximately like a blackbody at $T = 5800$ K, has a peak in B_λ at 5000 Å, which is the wavelength of green light, in the middle of the visual regime. In fact, the eyesight of most animals on Earth apparently evolved to have the most sensitivity in the wavelength range within which the Sun emits the most energy. (No less important, this wavelength range also coincides with the transmission range of water vapor in the atmosphere.) Note that the frequency ν_{max} where B_ν peaks is *not* the same as the frequency $\nu = c/\lambda_{max}$ at which B_λ peaks. The two spectral distributions are different, because a constant frequency interval $d\nu$ corresponds to a changing wavelength interval

$$d\lambda = \left|\frac{d\lambda}{d\nu}\right| d\nu = \frac{c}{\nu^2} d\nu \qquad (2.17)$$

that grows with wavelength (and falls with frequency).

Figure 2.1 shows blackbody spectral distributions for a variety of temperatures. The following features are important to note. First, the functions described above (u_ν, B_λ, etc.) are determined uniquely by one parameter, the temperature. Second, far from their peak frequencies or wavelengths, the Planck blackbody spectra assume two simple forms, as can be easily verified by taking the appropriate limits in Eqs. 2.4 and 2.12. At frequencies ν much lower than the peak (i.e., at photon energies $h\nu \ll kT$),

$$B_\nu \approx \frac{2\nu^2}{c^2} kT \qquad (2.18)$$

or

$$B_\lambda \approx 2ckT\lambda^{-4}. \qquad (2.19)$$

This is called the **Rayleigh-Jeans** side of the thermal spectrum. At frequencies much higher than the peak (photon energies $h\nu \gg kT$) the blackbody spectrum falls off exponentially with frequency as

$$B_\nu \sim e^{-(h\nu/kT)}, \qquad (2.20)$$

or with decreasing wavelength as

$$B_\lambda \sim e^{-(hc/\lambda kT)}. \qquad (2.21)$$

This is called the **Wien tail** of the distribution.

2.2 Measurement of Stellar Parameters

2.2.1 Distance

Distances to the nearest stars can be measured via **trigonometric parallax**. With current technology, about 100,000 stars have had their distances measured in this way. The motion

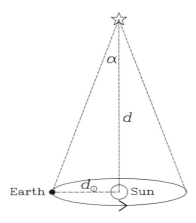

Figure 2.3 Schematic view of the apparent parallax motion of a nearby star, situated in the direction above the ecliptic plane, due to the Earth's circular orbit around the Sun.

of the Earth around the Sun produces an apparent movement on the sky of nearby stars, relative to more distant stars. Stars in the direction perpendicular to the plane of the Earth's orbit (called the *ecliptic plane*) will trace a circle on the sky in the course of a year (see Fig. 2.3), whereas stars in the directions of the ecliptic plane will trace on the sky a line segment that doubles back on itself. In other directions, stars will trace out an ellipse. The angular size of the semi-major axis of the ellipse will obviously be

$$\alpha = \frac{d_\odot}{d},$$ (2.22)

with $d_\odot$ the Earth–Sun distance and d the distance to the star. (The subscript $\odot$ marks properties of the Sun—distance, mass, radius, etc.). The distance $d_\odot$, which is also referred to as 1 astronomical unit (AU), is about 1.5×10^8 km. Parallax is actually used to define another unit of length, a **parsec** (pc). One pc is defined as the distance for which the parallax of 1 AU is 1 arcsecond (i.e., 1/3600 of a degree of arc, or $\pi/(180 \times 3600)$ radian). Thus,

$$1 \text{ pc} = 2.1 \times 10^5 \text{ AU} = 3.1 \times 10^{18} \text{ cm} = 3.3 \text{ ly},$$ (2.23)

where we have also expressed a parsec in light years (ly), the distance light travels in vacuum during a year. A light year is

$$1 \text{ ly} = 365.25 \times 24 \times 3600 \times c = 3.15 \times 10^7 \text{ s} \times 3 \times 10^{10} \text{ cm s}^{-1}.$$ (2.24)

It is convenient to remember that the number of seconds in a year is (by pure coincidence) close to $\pi \times 10^7$. The few nearest stars to the Solar System have distances of about 1 pc. Most of the stars visible to the naked eye are closer than 100 pc. At larger distances, convenient units are the kiloparsec (kpc; 10^3 pc), megaparsec (Mpc; 10^6 pc), and gigaparsec (Gpc; 10^9 pc).

Apart from the apparent motion of stars due to parallax, stars have real motions relative to each other, and hence relative to the Sun. Over human timescales, these real relative motions will generally appear on the sky to have constant velocity and direction. In practice, therefore, the parallax motion of nearby stars will often be superimposed on a linear **proper motion**, producing a curly or wavy trajectory on the sky.

2.2.2 Stellar Temperatures and Stellar Types

As we will see later on, the volume of every star has a range of temperatures, from millions of degrees Kelvin in its core to only thousands in the outer regions. However, the emitted spectrum of a star is largely determined by the temperature in the outermost "surface," or, more correctly, in its **photosphere**. The photosphere can be roughly defined as the region from which photons are able to escape a star without further absorption or scattering (see Fig. 2.4). As we will see, the scattering and absorption probabilities can have a strong dependence on the wavelength of the photon, and therefore the depth of the base of the photosphere can be wavelength dependent.

Material at the base of the photosphere emits approximately a Planck spectrum, which is then somewhat modified by the cooler, partly transparent, gas above it. By examining the emerging spectrum, one can then define various temperatures. The **color temperature** is the temperature of the Planck function with shape most closely matching the observed spectrum. For example, if we could identify the position of the peak of the spectrum, we could use Wien's law to set the temperature. In practice, the peak will often be outside the wavelength range for which we have data, and furthermore it is a broad feature that is hard to identify, especially given the modifications by cold absorbing gas above the last scattering surface.

A more practical variant is to measure the ratio of fluxes at two different wavelengths, $f_\lambda(\lambda_1)/f_\lambda(\lambda_2)$, and to find the temperature of the blackbody that gives such a ratio. Such a ratio is, in effect, what one always means by color. For example, when we say that the light from an object (whether intrinsic or reflected) appears, say, "red," we mean that we are detecting a larger ratio of red photons to blue photons than we would from an object that we would call "white." Color temperature can, of course, also be found by fitting the Planck spectrum to measurements at more than two wavelengths. Note that a color temperature cannot be found if all the measurements are well on the Rayleigh-Jeans (i.e., low-energy) side of the distribution. On that side the Planck spectrum has the same shape for all temperatures ($f_\lambda \propto \lambda^{-4}$ or $f_\nu \propto \nu^2$), and hence the ratio of fluxes at two wavelengths or frequencies will be the same, irrespective of temperature. In such a case we can only deduce that all our measurements are on the Rayleigh-Jeans side of a Planck spectrum, and we can set a lower limit on the temperature of the spectrum (see Problem 2).

Another kind of temperature can be associated with the photosphere of a star, by examining the absorption features at discrete wavelengths in the stellar spectrum. These absorptions are induced by atoms and molecules in the cooler, less dense, gas above the surface of last scattering. Photons with energies equal to those of individual quantum energy

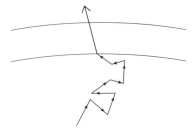

Figure 2.4 A photon inside a star is scattered many times until it reaches a radius from which it can escape. The last scattering surface defines the base of the photosphere of the star.

transitions of those atoms and molecules will be preferentially absorbed, and therefore depleted, from the light emerging from the photosphere of the star in the direction of a distant observer. The same atom or molecule, which will be excited to a higher energy level by absorbing a photon, can eventually decay radiatively and reemit a photon of the same energy. However, the reemitted photon will have a random direction, which will generally be different from the original direction toward the observer. Furthermore, the atom can undergo collisional deexcitation, in which it transfers its excitation energy to the other particles in the gas.

The wavelengths and strengths of the main absorption features, or **absorption lines** as they are often called, are primarily dependent on the level of ionization and excitation of the gas. The form of the absorption spectrum therefore reflects mainly the temperature of the photosphere, and only slightly the photosphere's chemical composition, which is actually similar in most stars.

It is of particular relevance to recall the quantum structure of the hydrogen atom. Hydrogen is the simplest atom, and it is therefore useful for understanding how stellar absorption spectra are produced. Furthermore, most stars are composed primarily of hydrogen; 92% of the atoms, or 75% of the mass, is hydrogen. Almost all of the rest is helium, and the heavier elements contribute only trace amounts. In fact, this elemental makeup is typical of almost all astronomical objects and environments, other than rocky planets like the Earth and some special types of stars. The heavier elements, despite their low abundances in stars, still play important physical and observational roles, as we will see later on.

The nth energy level of the hydrogen atom ($n = 1$ is the ground state) is given by the Bohr formula,

$$E_n = -\frac{e^4 m_e}{2\hbar^2} \frac{1}{n^2} = -13.6 \text{ eV} \frac{1}{n^2}, \tag{2.25}$$

where e is the electron charge in cgs units (e.s.u.), m_e is the electron mass, and $\hbar$ is Planck's constant divided by 2π. As one goes to higher n, the energy levels become more

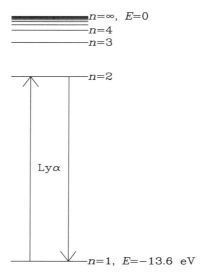

Figure 2.5 Energy levels of the hydrogen atom. Arrows indicate excitation from the ground state ($n = 1$) to the first excited energy level ($n = 2$), and deexcitation back to the ground state. Such excitation and deexcitation could be caused by, e.g., absorption by the atom of a Lyman-α photon, and subsequent spontaneous emission of a Lyman-α photon.

and more crowded (see Fig. 2.5). When $n \to \infty$, 13.6 eV above the ground state, we reach the "continuum" where $E_n = 0$ and the electron is free, i.e., the atom is ionized. Thus, absorption of an energy quantum of 13.6 eV, or more, can ionize a hydrogen atom that is in its ground state. The energy difference between two levels is

$$E_{n_1, n_2} = 13.6 \text{ eV} \left(\frac{1}{n_1^2} - \frac{1}{n_2^2} \right). \tag{2.26}$$

The wavelength[1] of a photon emitted or absorbed in a radiative transition between two levels will be

$$\lambda_{n_1, n_2} = \frac{hc}{E_{n_1, n_2}} = \frac{911.5 \text{ Å}}{1/n_1^2 - 1/n_2^2}. \tag{2.27}$$

A glass tube in the laboratory filled with atomic hydrogen at low pressure, when excited by an electrical discharge, will radiate photons at the discrete wavelengths corresponding to all of these electronic energy transitions. Such a photon spectrum is often called an *emission-line spectrum*, and is distinct from the thermal (i.e., blackbody) spectrum emitted by dense matter.

It is customary to group the different energy transitions of atomic hydrogen by a name identifying the lower energy level involved in the transition, combined with a Greek letter

---

[1] As customary in the astronomical research literature, wavelengths of atomic transitions are cited to four significant digits, as measured *in air* at standard temperature and pressure. Since the speed of light is smaller in air than in vacuum, the wavelengths in vacuum are longer by a factor equal to the index of refraction of air, 1.00028 for optical light.

that indicates the upper level of the transition. Thus, the **Lyman series** consists of all transitions to the $n = 1$ ground level:

Lyα : $2 \leftrightarrow 1, 1216$ Å

Lyβ : $3 \leftrightarrow 1, 1025$ Å

Lyγ : $4 \leftrightarrow 1, 972$ Å

etc.,

up until the *Lyman continuum*,

Ly$_{con}$: $\infty \leftrightarrow 1, <911.5$ Å.

Similarly, the **Balmer series** includes all transitions between the $n = 2$ state and higher states:

Hα : $3 \leftrightarrow 2, 6563$ Å

Hβ : $4 \leftrightarrow 2, 4861$ Å

Hγ : $5 \leftrightarrow 2, 4340$ Å

etc.,

up until the *Balmer continuum*,

Ba$_{con}$: $\infty \leftrightarrow 2, <3646$ Å.

In the same way, the Paschen series, Brackett series, and Pfund series designate transitions where $n = 3$, $n = 4$, and $n = 5$, respectively, are the lower levels. The photon wavelengths of the Lyman series are in the ultraviolet (UV) region of the electromagnetic spectrum, and the Paschen, and higher, series occur at infrared (IR), and longer, wavelengths. The Balmer series is of particular interest to us here, as it occurs in the optical region of the spectrum, where Earth's atmosphere has a transmission window. The atmosphere is almost completely opaque to photons of wavelengths shorter than ≈ 3100 Å, from ultraviolet through X-rays and γ-rays. At the infrared wavelengths longer than 10,000 Å (1 μm), there are only a few transmission "troughs," until one gets to millimeter (called *microwave*) wavelengths and longer, where the atmosphere is again transparent to radio-frequency electromagnetic radiation.

Early in the 20th century, before stellar physics was understood, stars were classified into a series of **spectral types** according to the types and strengths of the absorption lines appearing in their optical spectra. Figure 2.6 shows examples covering the range of spectral properties of most stars. Let us begin with *A-type* stars, the third from the top in the sequence shown (the meanings of the "5" and of the "V" after the "A" are explained below, and at the end of this chapter, respectively). Absorption in the hydrogen Balmer series is the most conspicuous feature in A-star spectra, starting with Hα at 6563 Å, proceeding up the series to shorter wavelengths, and to the sharp drop at the wavelength of the Balmer continuum at 3646 Å. Moving up in the figure to *B-type* stars, the hydrogen lines become weaker, and some other lines, due to absorption by helium, appear. At the top of the sequence, *O stars* have only very weak hydrogen Balmer lines, and some additional weak lines due to singly ionized helium. Working back down along the sequence, in the so-called F-type stars, the Balmer lines are again weaker than in A stars, but additional lines appear,

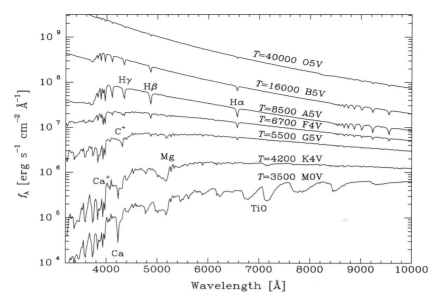

Figure 2.6 Zoom-in on the optical wavelength region of the stellar spectra shown in Fig. 2.1. The curves are labeled with their spectral types, in addition to the corresponding blackbody temperatures, which constitute the *effective temperatures* of the stars. Note the various labeled absorption features that appear and disappear as one goes from one spectral type to another. Data credit: R. Kurucz.

and are due to transitions in neutral and singly ionized light metals, mainly calcium, magnesium, and sodium. Progressing to G stars, the Balmer lines weaken further, while the absorptions due to metals become stronger. This trend continues in K stars where, in addition, molecular "bands" begin to appear. Such bands are actually numerous adjacent absorptions due to individual rotational, vibrational, and electronic quantum transitions of particular molecules, which have merged into broad absorption troughs. Finally, in M stars, at the bottom of the sequence, the molecular bands, notably due to TiO (titanium oxide), become prominent. The absorption bands seen in the M-star spectrum between 6600 and 8600 Å are mainly due to this molecule.

The names assigned to the different spectral types have a historical origin and no physical significance. However, the order in which they appear in Fig. 2.6 is one of temperature, with the hottest stars at the top and the coolest at the bottom. This is apparent at once by considering the colors of the stars. The O and B stars are clearly blue, with their spectra rising toward the shorter UV wavelengths of the peak of the blackbody spectrum. The A, F, and G stars become progressively "whiter." The K and M stars are clearly red, with their peak emission shifted to the infrared. The optical wavelength range shown probes the beginning of the Wien tail of the Planck function approximating these red stars.

Once the stars are ordered by temperature, the different absorption-line spectra can be understood as arising simply from the differences in the temperatures of the photospheres. In the cooler stars, the hydrogen atoms, which make up most of the cool gas above the last scattering surface, are almost all in their ground states. Among the photons in the emerging beam, only Lyman-series photons, in the UV range, can therefore be absorbed out by the hydrogen atoms, while optical photons emerge through the photosphere unhindered by the hydrogen. The metal atoms, though they exist only in trace amounts, do have energy-level transitions corresponding to optical-wavelength photons, and hence these atoms have high probabilities for absorbing photons. Thus, they leave a strong imprint on the spectrum, despite their rarity compared with hydrogen.

Going to warmer stars, frequent collisions between atoms in the photosphere cause a nonnegligible fraction of the hydrogen atoms at any given time to be in the first excited ($n = 2$) energy state. Now the optical photons with the Balmer-transition energies can be absorbed out of the energing light by the hydrogen atoms, thereby exciting the hydrogen atoms to higher levels, or ionizing them when the photon wavelengths are less than 3646 Å. In the hottest stars, the temperature in the photosphere is high enough that almost all of the hydrogen is ionized (as are the metal atoms). The photosphere then becomes transparent again to photons with the energies of the Balmer transitions, which have a low probability of being absorbed. Models of stellar atmospheres can be calculated, taking into account the detailed atomic physics and the passage of radiation through the gas, for a range of physical conditions, specifically the temperature. (In fact, the stellar spectra shown in Figs. 2.1 and 2.6 are theoretical models calculated by Kurucz.) Such theoretically calculated absorption spectra can be compared to the actual spectrum of a given stellar type, and thus the photospheric temperature can be accurately determined.

Generations of astronomy students have memorized the names of the stellar spectral types, ordered by decreasing temperature, with the mnemonic: "Oh Be A Fine Girl/Guy, Kiss Me!" There is a continuous transition in spectral properties between types, and astronomers quantify this by assigning, after the letter, a number between 0 and 9, with a larger number indicating a lower temperature. The Sun is a G2 star, and its spectrum is largely indistinguishable from that of any other normal star of this type. As we will see, all of the main physical properties (mass, radius, luminosity) of the stars sharing a common spectral classification are the same. For completeness, we note that the spectral sequence extends beyond M stars to two cooler classes, labeled L and T. Strictly speaking, members of these classes are not stars but *brown dwarfs*, objects intermediate between stars and giant planets in their properties. We return to brown dwarfs in section 4.2.3.4.

2.2.3 Luminosity and Radius

For a star with known distance and measured flux, the luminosity is

$$L = f 4\pi d^2. \tag{2.28}$$

This luminosity, integrated over all wavelengths, is called the **bolometric luminosity**. If the temperature of the stellar photosphere is known, one can then derive the stellar radius, r_*, from

$$L = 4\pi r_*^2 \sigma T^4. \tag{2.29}$$

Alternatively, if L and r_* are known and one determines a temperature from this relation, then this temperature is called the **effective temperature**, T_E. The radius of the Sun is

$r_\odot = 7.0 \times 10^{10}$ cm. As will be explained in more detail in section 2.2.4, below, stars that are members of a particular type of binary system, called *double-lined spectroscopic eclipsing binaries*, can have their radii measured. Of order 100 stars currently have such radius measurements, which are accurate to a few percent or better. The radii of a few other nearby stars have been measured, in some cases to better than 1% accuracy, using interferometric observations.

2.2.4 Binary Systems and Measurements of Mass

A direct measurement of stellar mass is generally possible only in certain **binary** (i.e., double) or multiple star systems. A significant fraction of all stars are members of binary systems. (The Sun is likely an example of a single star.) Observationally, binary systems are classified into various types. **Visual binaries** are pairs of stars in which both members are resolved individually, and may be seen orbiting their common center of mass. In most cases, the separation between the members is so large that the orbital period is very long by human timescales. In **astrometric binaries**, one observes the minute periodic motion on the sky of one member, as it orbits the system's common center of mass, even if the companion is too faint to be seen. In **eclipsing binaries**, the orbital plane of a pair (which, in general, is spatially unresolved) is inclined enough (i.e., close enough to edge-on) to our line of sight that each of the members periodically eclipses the other. The presence of a binary will then be revealed if the light from the system is monitored as a function of time. During each orbital period, the brightness of the system will undergo two "dips" (see Fig. 2.7), each corresponding to the eclipse of one star by the other. The depth of the dips will depend on the relative sizes and luminosities of the two stars.

A **spectroscopic binary** is a spatially unresolved pair that is revealed as a binary by its spectrum. For example, the observed photospheric absorption spectrum may be the super-position of the spectra of two different types of stars. Alternatively, even if the members are of the same type, their orbital velocities, v, may cause large enough Doppler shifts, $\Delta\lambda/\lambda = v/c$, in the wavelengths of absorption, to produce distinct lines, with shifts that oscillate periodically during each orbit (see Fig. 2.8). Sometimes, one of the members may be too faint, or devoid of strong absorption lines, to be detectable in the combined spectrum, but its presence will still be revealed by the periodically changing Doppler shifts of the brighter star. In fact, this is the very method by which planets orbiting other stars have been discovered in recent years (see below).

To see how binaries sometimes allow stellar mass determination, let us review some aspects of the Keplerian two-body problem, in which two masses orbit their common center of mass in elliptical trajectories. For simplicity, let us consider only circular orbits. The center of mass of two spherical masses is at the point between them where

$$r_1 M_1 = r_2 M_2, \tag{2.30}$$

with M_1 and M_2 being the masses and r_1 and r_2 their respective distances to the center of mass (see Fig. 2.9, left). Thus, if $a = r_1 + r_2$ is the separation between the masses,

$$r_1 = \frac{M_2}{M_1}(a - r_1) \tag{2.31}$$

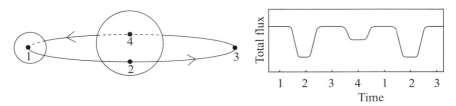

Figure 2.7 Schematic view of an eclipsing binary system (left), and its total brightness as a function of time (right). Numbers indicate the corresponding points on the orbit and in the so-called "light curve."

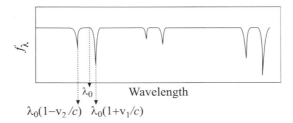

Figure 2.8 Schematic example of the spectrum of a double-lined spectroscopic binary. Each of the absorption lines in the spectrum appears twice, Doppler-shifted to longer and shorter wavelengths, respectively, as a result of the orbital motion of the binary members about their center of mass. During the orbital period, each absorption line oscillates back and forth about the restframe wavelength λ_0.

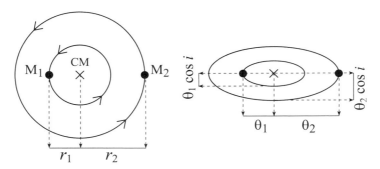

Figure 2.9 *Left:* A binary system, viewed pole-on, with its members in circular orbits with physical radii r_1 and r_2 around their common center of mass. *Right:* The appearance of the system when viewed as a visual binary, with the orbital plane inclined by an angle i to the line of sight, and orbital radii subtending angles on the sky θ_1 and θ_2. The circular orbits now appear as ellipses, with minor axes foreshortened by $\cos i$.

or

$$r_1 = \frac{M_2}{M_1 + M_2} a \qquad (2.32)$$

and

$$r_2 = \frac{M_1}{M_1 + M_2} a. \qquad (2.33)$$

Each of the masses is subject to the mutual gravitational attraction, and as a result orbits the center of mass with an angular frequency ω. The equation of motion for the first mass is then

$$M_1 \omega^2 r_1 = \frac{G M_1 M_2}{a^2}, \tag{2.34}$$

with G the gravitational constant. After substitution of r_1 from Eq. 2.32, this becomes Kepler's law:

$$\omega^2 = \frac{G(M_1 + M_2)}{a^3}. \tag{2.35}$$

A simple example in which Kepler's law can be used to determine a stellar mass is in the case of the Sun. The mass of the Earth is negligible compared to the Sun, so

$$M_\odot \approx \frac{\omega^2 a^3}{G} = \frac{4\pi^2 a^3}{\tau^2 G}, \tag{2.36}$$

where $\tau = 2\pi/\omega$ is the orbital period, i.e., 1 year. In cgs units the mass of the Sun is then

$$M_\odot = \frac{4 \times \pi^2 (1.5 \times 10^{13} \text{ cm})^3}{(3.15 \times 10^7 \text{ s})^2 \times 6.7 \times 10^{-8} \text{ erg cm g}^{-2}} = 2.0 \times 10^{33} \text{ g}. \tag{2.37}$$

In a visual binary, we can measure directly on the sky the angular separations θ_1 and θ_2 between each star and the common center of mass that they orbit. The perpendicular to the plane of the orbit will generally be inclined to our line of sight by some angle i, and as a result the circular orbits will appear projected on the sky as ellipses. If we can follow a good part of an entire orbit, this will not constitute a problem, as the semi-major axes of the ellipses will correspond to the angular radii of the deprojected circular orbits (see Fig. 2.9, right). Since both stars are at the same distance d from us, the ratio of the angles gives the ratio of the stellar masses:

$$\frac{\theta_1 d}{\theta_2 d} = \frac{r_1}{r_2} = \frac{M_2}{M_1}. \tag{2.38}$$

Given the distance (which allows deriving the physical separation a) and the observed period, Kepler's law yields $M_1 + M_2$. Together with Eq. 2.38, we can solve for M_1 and M_2 individually.

In spectroscopic binaries, we cannot measure directly the separations a, r_1, and r_2. Instead, we can use the amplitudes of the oscillations in line-of-sight velocities deduced from the Doppler shifts. Because the perpendicular to the orbital plane is inclined to the line of sight by an angle i (see Fig. 2.10), the Doppler velocity amplitudes we measure will be related to the true orbital velocity amplitudes by

$$|v_{1\text{obs}}| = |v_1| \sin i, \quad |v_{2\text{obs}}| = |v_2| \sin i. \tag{2.39}$$

But since

$$|v_1| = \frac{2\pi r_1}{\tau}, \quad |v_2| = \frac{2\pi r_2}{\tau}, \tag{2.40}$$

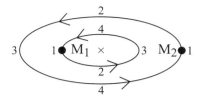
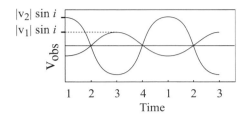

Figure 2.10 *Left:* A spectroscopic binary system with circular orbits and with orbital plane inclined by an angle i to the line of sight. *Right:* Observed velocity of each of the components, as deduced from the Doppler shift of its spectral features, as a function of time. Negative velocities are approaching and positive are receding. Numbers indicate the corresponding points on the orbits and in the so-called *radial-velocity curve.*

then

$$\frac{|v_{1\text{obs}}|}{|v_{2\text{obs}}|} = \frac{r_1}{r_2} = \frac{M_2}{M_1}. \tag{2.41}$$

Replacing, in Kepler's law, a with $r_1 + r_2$, and using Eqs. 2.39 and 2.40 to express r_1 and r_2, we obtain

$$(M_1 + M_2) \sin^3 i = \frac{\tau (|v_{1\text{obs}}| + |v_{2\text{obs}}|)^3}{2\pi G}. \tag{2.42}$$

We see that in spectroscopic binaries the inclination of the orbits is an additional unknown variable that enters the mass determination. In such systems, we will there-fore be able to determine the stellar masses only up to a factor $\sin^3 i$. An exception to this is the case of eclipsing spectroscopic binaries. There, the fact that the members of a pair eclipse each other implies i must be close to 90°, and the individual masses can therefore be found. Indeed, in such systems, one can use the detailed shape of the *light curve* of the eclipse (e.g., Fig. 2.7), combined with the known relative velocity of the two stars as they pass one in front of the other, to deduce the physical radii of the two stars (see Problem 5), as well as the precise value of i.

In many spectroscopic binaries, the spectrum of only one star is detected, due to the faintness of the secondary object, M_2. The presence of a companion is deduced solely from the periodic velocity oscillations in the spectral lines of one star, say, M_1. In this case, we can use Eq. 2.41 to express the unmeasured $|v_{2\text{obs}}|$, giving

$$(M_1 + M_2) \sin^3 i = \frac{\tau |v_{1\text{obs}}|^3 (1 + M_1/M_2)^3}{2\pi G} \tag{2.43}$$

or

$$\frac{M_2^3}{(M_1 + M_2)^2} \sin^3 i = \frac{\tau |v_{1\text{obs}}|^3}{2\pi G}, \tag{2.44}$$

and there is now only one equation for three unknowns, M_1, M_2, and $\sin i$.

An important case is when $M_2 \ll M_1$, where Eq. 2.44 simplifies to

$$M_2 \sin i \approx \left(\frac{\tau}{2\pi G}\right)^{1/3} |v_{1\text{obs}}| M_1^{2/3}. \tag{2.45}$$

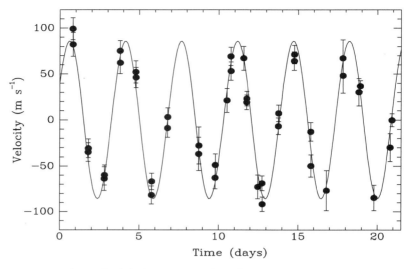

Figure 2.11 Observed *radial velocity curve* (i.e., line-of-sight velocity vs. time) for the nearby (≈50 pc) star HD209458, revealing the presence of a planetary companion that causes the star to orbit around the common center of mass. Using Eq. 2.45, and assuming the parent star has a mass $M_1 = 1.1 M_\odot$ (based on its observed spectral type), the amplitude (86 m s^{-1}) and period (3.5 days) of the observed velocity oscillations indicate a planet mass $M_2 = 0.7 M_J / \sin i$, where M_J is one Jupiter mass. The orbital radius is 0.05 AU. In this particular system, a periodic 1.5% amplitude eclipse of the star's flux has also been detected, due to the transit of the planet across the face of the star, indicating $\sin i \approx 1$. Data credit: T. Mazeh et al. 1999, *Astrophys. J.*, 532, L55.

This case applies to searches for **extrasolar planets**, by means of the small wobble a planet induces on its parent star. The parent star's mass, M_1, needs to be estimated by some other means, normally by identifying its spectral type. We can calculate from Eq. 2.45 the expected amplitude of the velocity oscillations of, say, a $1 M_\odot$ star that is orbited by a Jupiter-mass ($10^{-3} M_\odot$) planet at a radius of 1 AU:

$$|v_{1\mathrm{obs}}| \approx M_2 \sin i\, M_1^{-2/3} \left(\frac{\tau}{2\pi G}\right)^{-1/3}$$

$$= 10^{-3} \times 2 \times 10^{33}\ \mathrm{g} \times (2 \times 10^{33}\ \mathrm{g})^{-2/3} \left(\frac{3.15 \times 10^7\ \mathrm{s}}{2\pi \times 6.7 \times 10^{-8}\ \mathrm{cgs}}\right)^{-1/3}$$

$$= 31\ \mathrm{m\ s^{-1}}, \tag{2.46}$$

where we have assumed $\sin i \approx 1$. (Here and henceforth, we abbreviate with "cgs" the units of the gravitational constant G, erg cm gr^{-2}.) Such a velocity produces a tiny Doppler shift in the star's spectral lines, of $\Delta\lambda/\lambda = v/c = 10^{-7}$. Nevertheless, sensitive spectroscopic techniques have been developed for this purpose, and hundreds of extrasolar planets have been discovered, and their masses determined (in most cases to an unknown factor $\sin i$) in this way. Figure 2.11 shows an example. These extrasolar planets indeed have masses of order that of Jupiter, but are sometimes in orbits with periods of just a few days, indicating orbital radii of a few hundredths of an AU.

In reality, many binary systems are in elliptical, rather than circular orbits. Two additional parameters are then required to describe the problem—the eccentricity of the orbit and the orientation angle of the ellipse in the orbital plane, as viewed from our vantage point. However, these parameters can be determined from the data (e.g., from the shape of the orbit on the sky, or from the functional form of the radial–velocity curve of each component, or from asymmetries in the timing and duration of primary and secondary eclipses). The essential possibilities and limitations of stellar mass determination we have found above for various types of binaries, assuming circular orbits, hold also for binaries with elliptical orbits.

2.3 The Hertzsprung-Russell Diagram

A crucial step toward understanding stellar physics was taken in 1911 independently by two astronomers, Hertzsprung and Russell, who placed measurements of stars on a logarithmic plot with axes of luminosity and effective surface temperature.[2] Figure 2.12 shows an example of such an **H-R diagram**. Note that temperature is traditionally shown growing to the left on such a plot. Almost all known stars are concentrated in several well-defined loci on such a diagram. About 80–90% of all stars (depending on what stellar environment one looks at) lie in a narrow diagonal strip called the **main sequence**, which corresponds very roughly to a relation $L \sim T_E^8$. Since the luminosity of a spherical blackbody radiator is

$$L = 4\pi r_*^2 \sigma T_E^4, \tag{2.47}$$

this immediately implies that hotter stars are bigger, with $r_* \sim T_E^2$. The Sun is a main sequence star. Since the coolest stars have about half the surface temperature of the Sun and the hottest stars have about 5 times the solar temperature, the radii of main-sequence stars are in the range of about 1/4 to 25 times the solar radius.

Two other stellar loci are apparent on the H-R diagram. There is a concentration of points corresponding to stars that are cool (i.e., red) yet with luminosities orders of magnitudes higher than those of main sequence stars. Clearly, these must be objects with large surface areas, with radii of order 100 times the solar radius, i.e., $\sim 10^8$ km, or about 1 AU. Accordingly, such stars are named **red giants**. In the lower left part of the diagram there is another sequence of points, corresponding to stars that are quite hot (i.e., white to blue), yet with luminosities that are orders of magnitude smaller than those of main sequence stars with such temperatures. These must be stars with much smaller radii, of order 10^4 km (i.e., of the order of the Earth's radius, 6400 km), and they are therefore labeled **white dwarfs**.

In terms of physical meaning, it was originally speculated (incorrectly) that the main sequence is a cooling sequence, in which stars are born hot and then move along the sequence as they cool. (This led to the confusing terminology, sometimes still used today,

---

[2] Historically, what was first plotted was the fluxes from stars that are in a cluster, and hence all at the same distance, and their colors.

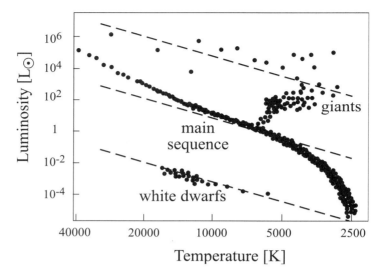

Figure 2.12 Schematic example of an H-R diagram, showing stellar luminosity vs. effective surface temperature. Note that the axes are logarithmic and that the temperature grows to the left. The dashed lines show the luminosity temperature relation, $L = 4\pi r_*^2 \sigma T_E^4$, for objects of constant radius. The three lines correspond, from top to bottom, to $r_* = 100 r_\odot$, $r_\odot$, and $0.01 r_\odot$.

of "early-type" stars for O and B spectral types and "late-type" stars for FGKM. Main-sequence stars are sometimes also called dwarfs, to distinguish them from giant stars, a potential source for additional confusion.) However, as mass measurements for stars that are members of binaries became available, it became clear that the main sequence is a **mass** sequence, with high-mass stars at high luminosities and high effective temperatures, and low-mass stars at low luminosities and low effective temperatures. The range of masses on the main sequence is between ∼0.1 and ∼100$M_\odot$. The luminosity as a function of mass goes as $L \sim M^\alpha$, with $\alpha \approx 3$ for stars more massive than the Sun and $\alpha \approx 5$ for the less massive stars. Red giants, despite their large luminosities, turn out to have masses usually in the range 1–2$M_\odot$. White dwarfs generally have masses similar to that of the Sun, or lower, and are never more massive than 1.4$M_\odot$.

As we will see, a star spends most of its lifetime at the same location on the main sequence. This is the period during which nuclear hydrogen burning takes place in the stellar core. The processes that occur, once most of the hydrogen in the core has been synthesized into helium, lead to a large, but short-lived, increase in luminosity and an expansion of the star's outer layers, which is observed as the giant-star phase of stellar evolution. Stars that begin their lives with less than about 8$M_\odot$ eventually shed their outer layers and become white dwarfs—compact remnants of the original stellar core. White dwarfs are devoid of energy-producing nuclear reactions, and slowly radiate away their fossil heat. Stars more massive than about 8$M_\odot$, after passing through the giant stage (where they are called red or blue **supergiants** because of their large luminosities— these are the stars near the top edge of the H-R diagram)—undergo a runaway process of gravitational core collapse that ends (at least in some cases) in a *supernova* explosion. The stellar remnants in these cases are *neutron stars* and *black holes*. Compared to white dwarfs, neutron stars are even hotter (by an order of magnitude) and more compact (by three orders of magnitude in radius), and hence less luminous (by two orders of magnitude). They are not generally plotted on an H-R diagram; only a handful of isolated neutron stars (i.e., neutron stars not in binary systems or not still inside the debris of the supernova explosions that created them) have been detected optically, owing to their extreme faintness.

A final note on nomenclature: apart from their spectral types (B0, A5, G2, etc.), stars are divided into luminosity classes, which are labeled by Roman numerals going from I to V, with decreasing luminosity. Classes I to IV are giants, and V designates main sequence stars. Thus, the Sun is a G2V star. White dwarfs have their own classification system, with spectral type names beginning with a capital D.

Problems

1. a. If the Sun subtends a solid angle Ω on the sky, and the flux from the Sun just above the Earth's atmosphere, integrated over all wavelengths, is $f(d_\odot)$, show that the flux at the solar photosphere is $\pi f(d_\odot)/\Omega$.

   b. The angular diameter of the Sun is 0.57 degree. Calculate the solid angle subtended by the Sun, in steradians.
   Answer: $\Omega = 7.8 \times 10^{-5}$.

   c. The solar flux at Earth is

   $$f(d_\odot) = 1.4 \times 10^6 \text{ erg s}^{-1} \text{ cm}^{-2} = 1.4 \text{ kW m}^{-2}.$$

   Use (b), and the Stefan-Boltzmann law, to derive the effective surface temperature of the Sun.
   Answer: $T_E = 5800$ K.

   d. Derive an expression for the surface temperature of the Sun, in terms only of its solid angle, its flux per unit wavelength $f_\lambda(\lambda_1)$ at Earth at one wavelength λ_1, and fundamental constants.

2. a. Show that, if the ratio of the blackbody fluxes from a star at two different frequencies (i.e., a color) is measured, then, in principle, the surface temperature of the star can be derived, even if the star's solid angle on the sky is unknown (e.g., if it is too distant to be spatially resolved, and its distance and surface area are both unknown).

   b. Explain why it will be hard, in practice, to derive the temperature measurement if both frequencies are on the Rayleigh-Jeans side of the blackbody curve, $h\nu \ll kT$.

   c. For the case that both measurements are on the Wien tail of the blackbody curve, $h\nu \gg kT$, derive a simple, approximate, expression for the temperature as a function of the two frequencies and of the flux ratio at the two frequencies.

    d. If, in addition to the flux ratio in (c), a parallax measurement and the total flux (integrated over all frequencies) at Earth are available, show that the star's radius can be derived.

3. If parallax can be measured with an accuracy of 0.01 arcsecond, and the mean density of stars in the solar neighborhood is 0.1 pc^{-3}, how many stars can have their distances measured via parallax?
   Answer: 4.2×10^5.

4. The maximal radial velocities measured for the two components of a spectroscopic binary are 100 and 200 km s^{-1}, with an orbital period of 2 days. The orbits are circular.
   a. Find the mass ratio of the two stars.
   b. Use Kepler's law (Eq. 2.42) to calculate the value of $M \sin^3 i$ for each star, where M is the mass and i is the inclination to the observer's line of sight of the perpendicular to the orbital plane.
      Answer: $3.7 M_\odot$ and $1.8 M_\odot$.
   c. Calculate the mean expectation value of the factor $\sin^3 i$, i.e., the mean value it would have among an ensemble of binaries with random inclinations. Find the masses of the two stars, if $\sin^3 i$ has its mean value.
      Hint: In spherical coordinates, (θ, ϕ), integrate over the solid angle of a sphere where the observer is in the direction of the z axis, with each solid angle element weighted by $\sin^3 \theta$.
      Answers: $3\pi/16 = 0.59$; $6.3 M_\odot$ and $3.1 M_\odot$.

5. In an eclipsing spectroscopic binary, the maximal radial velocities measured for the two components are 20 and 5 km s^{-1}. The orbit is circular, and the orbital period is $P = 5$ yr. It takes 0.3 day from the start of the eclipse to the main minimum, which then lasts 1 day.
   a. Find the mass of each star. Since the binary is of the eclipsing type, one can safely approximate $i \approx 90°$. Check to what degree the results are affected by small deviations from this angle, to convince yourself that this is a good approximation.
      Answers: $M_1 = 2.3 M_\odot$, $M_2 = 0.58 M_\odot$.
   b. Assume again $i = 90°$ and find the radius of each star. Is the result still insensitive to the exact value of i?
      Answers: $r_1 = 2.0 r_\odot$, $r_2 = 0.46 r_\odot$.

3 | Stellar Physics

In this chapter, we obtain a physical understanding of main-sequence stars and of their properties, as outlined in the previous chapter. The Sun is the nearest and best-studied star, and its properties provide useful standards to which other stars can be compared. For reference, let us summarize the measured parameters of the Sun. The Earth–Sun distance is

$$d_\odot = 1.5 \times 10^8 \text{ km} = 1 \text{ AU}. \tag{3.1}$$

The mass of the Sun is

$$M_\odot = 2.0 \times 10^{33} \text{ g}. \tag{3.2}$$

The radius of the Sun is

$$r_\odot = 7.0 \times 10^{10} \text{ cm}. \tag{3.3}$$

Using the mass and the radius, we can find the mean density of the Sun,

$$\bar{\rho} = \frac{M_\odot}{\frac{4}{3}\pi R_\odot^3} = \frac{3 \times 2 \times 10^{33} \text{ g}}{4\pi \times (7 \times 10^{10} \text{ cm})^3} = 1.4 \text{ g cm}^{-3}. \tag{3.4}$$

The Sun's mean density is thus not too different from that of liquid water. The bolometric Solar luminosity is

$$L_\odot = 3.8 \times 10^{33} \text{ erg s}^{-1}. \tag{3.5}$$

When divided by $4\pi d_\odot^2$, this gives the Solar flux above the Earth's atmosphere, sometimes called the solar constant:

$$f_\odot = 1.4 \times 10^6 \text{ erg s}^{-1} \text{ cm}^{-2} = 1.4 \text{ kW m}^{-2}. \tag{3.6}$$

The effective surface temperature is

$$T_{E\odot} = 5800 \text{ K}. \tag{3.7}$$

From Wien's law (Eq. 2.16), the typical energy of a solar photon is then 1.4 eV. When the energy flux is divided by this photon energy, the photon flux is

$$f_{\odot,\text{ph}} \approx \frac{1.4 \times 10^6 \text{ erg s}^{-1} \text{ cm}^{-2}}{1.4 \text{ eV} \times 1.6 \times 10^{-12} \text{ erg eV}^{-1}} = 6.3 \times 10^{17} \text{ s}^{-1} \text{ cm}^{-2} \tag{3.8}$$

(where we have converted between energy units using $1 \text{ eV} = 1.6 \times 10^{-12}$ erg).
From radioactive dating of Solar System bodies, the Sun's age is about

$$t_\odot = 4.5 \times 10^9 \text{ yr.} \tag{3.9}$$

Finally, from a solution of the stellar models that we will develop, the central density and temperature of the Sun are

$$\rho_c = 150 \text{ g cm}^{-3}, \tag{3.10}$$
$$T_c = 15 \times 10^6 \text{K.} \tag{3.11}$$

3.1 Hydrostatic Equilibrium and the Virial Theorem

A star is a sphere of gas that is held together by its self gravity, and is balanced against collapse by pressure gradients. To see this, let us calculate the **free-fall timescale** of the Sun, i.e., the time it would take to collapse to a point, if there were no pressure support. Consider a mass element dm at rest in the Sun at a radius r_0. Its potential energy is

$$dU = -\frac{GM(r_0)dm}{r_0}, \tag{3.12}$$

where $M(r_0)$ is the mass interior to r_0. From conservation of energy, the velocity of the element as it falls toward the center is

$$\frac{1}{2} \left(\frac{dr}{dt} \right)^2 = \frac{GM(r_0)}{r} - \frac{GM(r_0)}{r_0}, \tag{3.13}$$

where we have assumed that the amount of (also-falling) mass interior to r_0 remains constant. Separating the variables and integrating, we find

$$\tau_{\text{ff}} = \int_0^{\tau_{\text{ff}}} dt = -\int_{r_0}^0 \left[2GM(r_0) \left(\frac{1}{r} - \frac{1}{r_0} \right) \right]^{-1/2} dr$$

$$= \left(\frac{r_0^3}{2GM(r_0)} \right)^{1/2} \int_0^1 \left(\frac{x}{1-x} \right)^{1/2} dx. \tag{3.14}$$

The definite integral on the right equals $\pi/2$, and the ratio $M(r_0)/r_0^3$, up to a factor $4\pi/3$, is the mean density $\bar{\rho}$, so

$$\tau_{\text{ff}} = \left(\frac{3\pi}{32G\bar{\rho}} \right)^{1/2}. \tag{3.15}$$

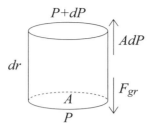

Figure 3.1 Hydrostatic equilibrium. The gravitational force F_{gr} on a mass element of cross-sectional area A is balanced by the force AdP due to the pressure difference between the top and the bottom of the mass element.

For the parameters of the Sun, we obtain

$$\tau_{\mathrm{ff}\odot} = \left(\frac{3\pi}{32 \times 6.7 \times 10^{-8} \ \mathrm{cgs} \times 1.4 \ \mathrm{g \ cm^{-3}}} \right)^{1/2} = 1800 \ \mathrm{s}. \tag{3.16}$$

Thus, without pressure support, the Sun would collapse to a point within half an hour.

This has not happened because the Sun is in **hydrostatic equilibrium**. Consider now a small, cylinder-shaped mass element inside a star, with A the area of the cylinder's base, and dr its height (see Fig. 3.1). If there is a pressure difference dP between the top of the cylinder and its bottom, this will lead to a net force Adp on the mass element, in addition to the force of gravity toward the center. Equilibrium will exist if

$$-\frac{GM(r)dm}{r^2} - AdP = 0. \tag{3.17}$$

But

$$dm = \rho(r)Adr, \tag{3.18}$$

which leads us to the first equation of stellar structure, the **equation of hydrostatic equilibrium**:

$$\boxed{\frac{dP(r)}{dr} = -\frac{GM(r)\rho(r)}{r^2}. \tag{3.19}}$$

Naturally, the pressure gradient is negative, because to counteract gravity, the pressure must decrease outward (i.e., with increasing radius.)

This simple equation, combined with some thermodynamics, can already provide valuable insight. Let us multiply both sides of Eq. 3.19 by $4\pi r^3 dr$ and integrate from $r = 0$ to r_*, the outer radius of the star:

$$\int_0^{r_*} 4\pi r^3 \frac{dP}{dr} dr = - \int_0^{r_*} \frac{GM(r)\rho(r)4\pi r^2 dr}{r}. \tag{3.20}$$

The right-hand side, in the form we have written it, is seen to be the energy that would be gained in constructing the star from the inside out, bringing from infinity shell by shell (each shell with a mass $dM(r) = \rho(r)4\pi r^2 dr$). This is just the gravitational potential self-energy of the star, E_{gr}. On the left side, integration by parts gives

$$[P(r)4\pi r^3]_0^{r_*} - 3\int_0^{r_*} P(r)4\pi r^2 dr. \tag{3.21}$$

We will define the surface of the star as the radius at which the pressure goes to zero. The first term is therefore zero. The second term is seen to be -3 times the volume-averaged pressure, $\bar{P}$, up to division by the volume V of the star. Equating the two sides, we obtain

$$\bar{P} = -\frac{1}{3}\frac{E_{gr}}{V}. \tag{3.22}$$

In words, the mean pressure in a star equals minus one-third of its gravitational energy density. Equation 3.22 is one form of the so-called **virial theorem** for a gravitationally bound system.

To see what Eq. 3.22 implies, consider a star composed of a classical, mono-atomic, nonrelativistic, ideal gas of N identical particles.[1] At every point in the star the gas equation of state is

$$PV = NkT, \tag{3.23}$$

and its thermal energy is

$$E_{th} = \tfrac{3}{2}NkT. \tag{3.24}$$

Thus,

$$P = \frac{2}{3}\frac{E_{th}}{V}, \tag{3.25}$$

i.e., the local pressure equals 2/3 the local thermal energy density. Multiplying both sides by $4\pi r^2$ and integrating over the volume of the star, we find that

$$\bar{P}V = \tfrac{2}{3}E_{th}^{tot}, \tag{3.26}$$

with E_{th}^{tot} the total thermal energy of the star. Substituting from Eq. 3.22, we obtain

$$E_{th}^{tot} = -\frac{E_{gr}}{2}, \tag{3.27}$$

---

[1] *Classical* means that the typical separations between particles are larger than the de Broglie wavelengths of the particles, $\lambda = h/p$, where p is the momentum. *Nonrelativistic* means that the particle velocities obey $v \ll c$. An ideal gas is defined as a gas in which particles experience only short-range (compared to their typical separations) interactions with each other—billiard balls on a pool table are the usual analog.

which is another form of the virial theorem. Equation 3.22 says that when a star contracts and loses energy, i.e., its gravitational self-energy becomes more negative, then its thermal energy rises. This means that stars have negative heat capacity—their temperatures rise when they lose energy. As we will see, this remarkable fact is at the crux of all of stellar evolution.

A third form of the virial theorem is obtained by considering the total energy of a star, both gravitational and thermal,

$$E_{\text{total}} = E_{\text{th}}^{\text{tot}} + E_{\text{gr}} = -E_{\text{th}}^{\text{tot}} = \frac{E_{\text{gr}}}{2}. \tag{3.28}$$

Thus, the total energy of a star that is composed of a classical, nonrelativistic, ideal gas is negative, meaning the star is bound. (To see what happens in the case of a relativistic gas, solve Problem 1.) Since all stars constantly radiate away their energy (and hence E_{total} becomes more negative), they are doomed to collapse (E_{gr} becomes more negative), eventually. We will see in chapter 4 that an exception to this occurs when the stellar gas moves from the classical to the quantum regime.

We can also use Eq. 3.22 to get an idea of the typical pressure and temperature inside a star, as follows. The right-hand side of Eq. 3.20 permits evaluating E_{gr} for a choice of $\rho(r)$. For example, for a constant density profile, $\rho = \text{const.}$,

$$E_{\text{gr}} = -\int_0^{r_*} \frac{GM(r)\rho(r)4\pi r^2 dr}{r} = -\int_0^{r_*} \frac{G\frac{4\pi}{3}r^3\rho^2 4\pi r^2 dr}{r} = -\frac{3}{5}\frac{GM_*^2}{r_*}. \tag{3.29}$$

A density profile, $\rho(r)$, that falls with radius will give a somewhat more negative value of E_{gr}. Taking a characteristic $E_{\text{gr}} \sim -GM^2/r$, we find that the mean pressure in the Sun is

$$\bar{P}_\odot \sim \frac{1}{3}\frac{GM_\odot^2}{\frac{4}{3}\pi r_\odot^3 r_\odot} = \frac{GM_\odot^2}{4\pi r_\odot^4} \approx 10^{15} \text{ dyne cm}^{-2} = 10^9 \text{atm}. \tag{3.30}$$

To find a typical temperature, which we will call the **virial temperature**, let us assume again a classical nonrelativistic ideal gas, with particles of mean mass $\bar{m}$. Equation 3.27 then applies, and

$$\frac{3}{2}NkT_{\text{vir}} \sim \frac{1}{2}\frac{GM_\odot^2}{r_\odot} = \frac{1}{2}\frac{GM_\odot N\bar{m}}{r_\odot}. \tag{3.31}$$

The mass of an electron is negligibly small, only $\approx 1/2000$ compared to the mass of a proton. For an ionized hydrogen gas, consisting of an equal number of protons and electrons, the mean mass $\bar{m}$,

$$\bar{m} = \frac{m_e + m_p}{2} = \frac{m_H}{2}, \tag{3.32}$$

is therefore close to one-half the mass of the proton or exactly one-half of the hydrogen atom, $m_H = 1.7 \times 10^{-24}$ g. The typical thermal energy is then

$$kT_{\rm vir} \sim \frac{GM_\odot m_H}{6r_\odot} = \frac{6.7 \times 10^{-8}\ {\rm cgs} \times 2 \times 10^{33}\ {\rm g} \times 1.7 \times 10^{-24}\ {\rm g}}{6 \times 7 \times 10^{10}\ {\rm cm}}$$

$$= 5.4 \times 10^{-10}\ {\rm erg} = 0.34\ {\rm keV}. \tag{3.33}$$

With $k = 1.4 \times 10^{-16}$ erg K^{-1} = 8.6×10^{-5} eV K^{-1}, this gives a virial temperature of about 4×10^6 K. As we will see, at temperatures of this order of magnitude, nuclear reactions can take place, and thus replenish the thermal energy that the star radiates away, halting the gravitational collapse (if only temporarily).

Of course, in reality, just like $P(r)$, the density $\rho(r)$ and the temperature $T(r)$ are also functions of radius and they grow toward the center of a star. To find them, we need to define additional equations. We will see that the equation of hydrostatic equilibrium is one of four coupled differential equations that determine stellar structure.

3.2 Mass Continuity

In the hydrostatic equilibrium equation (Eq. 3.19), we have $M(r)$ and $\rho(r)$, which are easily related to each other:

$$dM(r) = \rho(r)4\pi r^2 dr, \tag{3.34}$$

or

$$\frac{dM(r)}{dr} = 4\pi r^2 \rho(r). \tag{3.35}$$

Although this is, in essence, merely the definition of density, in the context of stellar structure this equation is often referred to as the **equation of mass continuity** or the **equation of mass conservation**.

3.3 Radiative Energy Transport

The radial gradient in $P(r)$ that supports a star is produced by a gradient in $\rho(r)$ and $T(r)$. In much of the volume of most stars, $T(r)$ is determined by the rate at which radiative energy flows in and out through every radius, i.e., the luminosity $L(r)$. To find the equation that determines $T(r)$, we need to study some of the basics of *radiative transfer*, the passage of radiation through matter. In some of the volume of some stars, the energy transport mechanism that dominates is convection, rather than radiative transport. We discuss convection is section 3.12. Energy transport by means of conduction plays a role only in dense stellar remnants—white dwarfs and neutron stars—which are discussed in chapter 4.

Photons in stars can be absorbed or scattered out of a beam via interactions with molecules, with atoms (either neutral or ions), and with electrons. If a photon traverses a

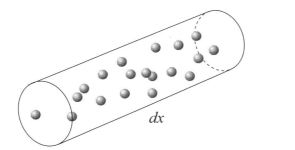

Figure 3.2 A volume element in a field of targets as viewed in perspective (left). The target number density is n, and each target presents a cross section σ. From the base of the cylindrical volume, $n\,dx$ targets per unit area are seen in projection (right). A straight line along the length of the volume will therefore intercept, on average, $n\sigma\,dx$ targets.

path dx filled with "targets" with a number density n (i.e., the number of targets per unit volume), then the projected number of targets per unit area lying in the path of the photon is $n\,dx$ (see Fig. 3.2). If each target poses an effective **cross section**[2] σ for absorption or scattering, then the fraction of the area covered by targets is $\sigma n\,dx$. Thus, the number of targets that will typically be intersected by a straight line traversing the path dx, or, in other words, the number of interactions the photon undergoes, will be

$$\text{\# of interactions} = n\sigma\,dx. \tag{3.36}$$

Equation 3.36 defines the concept of cross section. (Cross section can be defined equivalently as the ratio between the interaction rate per target particle and the incoming flux of projectiles.) Setting the left-hand side equal to 1, the typical distance a photon will travel between interactions is called the **mean free path**:

$$l = \frac{1}{n\sigma}. \tag{3.37}$$

More generally, the stellar matter will consist of a variety of absorbers and scatterers, each with its own density n_i and cross section σ_i. Thus,

$$l = \frac{1}{\sum n_i \sigma_i} \equiv \frac{1}{\rho\kappa}, \tag{3.38}$$

where we have used the fact that all the particle densities will be proportional to the mass density ρ, to define the **opacity** κ. The opacity obviously has cgs units of $\text{cm}^2\,\text{g}^{-1}$, and will depend on the local density, temperature, and element abundance.

We will return later to the various processes that produce opacity. However, to get an idea of the magnitude of the scattering process, let us consider one of the important interactions—**Thomson scattering** of photons on free electrons (see Fig. 3.3).

---

[2] The cross section of a particle, which has units of area, quantifies the degree to which the particle is liable to take part in a particular interaction (e.g., a collision or a reaction) with some other particle.

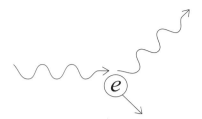

Figure 3.3 Thomson scattering of a photon on a free electron.

The Thomson cross section is

$$\sigma_T = \frac{8\pi}{3} \left(\frac{e^2}{m_e c^2} \right)^2 = 6.7 \times 10^{-25} \text{ cm}^2. \tag{3.39}$$

It is independent of temperature and photon energy.[3] In the hot interiors of stars, the gas is fully ionized and therefore free electrons are abundant. If we approximate, for simplicity, that the gas is all hydrogen, then there is one electron per atom of mass m_H, and

$$n_e \approx \frac{\rho}{m_H}. \tag{3.40}$$

The mean free path for electron scattering is then

$$l_{\text{es}} = \frac{1}{n_e \sigma_T} \approx \frac{m_H}{\rho \, \sigma_T} \approx \frac{1.7 \times 10^{-24} \text{ g}}{1.4 \text{ g cm}^{-3} \times 6.7 \times 10^{-25} \text{ cm}^2} \approx 2 \text{ cm}, \tag{3.41}$$

where we have used the mean mass density of the Sun calculated previously. In reality, the density of the Sun is higher than average in regions where electron scattering is the dominant source of opacity, while in other regions other processes, apart from electron scattering, are important. As a result, the actual typical photon mean free path is even smaller, and is $l \approx 1$ mm.

Thus, photons can travel only a tiny distance inside the Sun before being scattered or absorbed and reemitted in a new direction. Since the new direction is random, the emergence of photons from the Sun is necessarily a **random walk** process. The vector **D** describing the change in position of a photon after N steps, each described by a vector $\mathbf{l}_i$ having length l and random orientation (see Fig. 3.4), is

$$\mathbf{D} = \mathbf{l}_1 + \mathbf{l}_2 + \mathbf{l}_3 + \cdots + \mathbf{l}_N. \tag{3.42}$$

The square of the linear distance covered is

$$\mathbf{D}^2 = |\mathbf{l}_1|^2 + |\mathbf{l}_2|^2 + \cdots + |\mathbf{l}_N|^2 + 2(\mathbf{l}_1 \cdot \mathbf{l}_2 + \mathbf{l}_1 \cdot \mathbf{l}_3 + \cdots), \tag{3.43}$$

and its expectation value is

$$\langle \mathbf{D}^2 \rangle = N l^2. \tag{3.44}$$

---

[3] Note the inverse square dependence of the Thomson cross section on the electron mass. For this reason, protons and nuclei, which are much heavier, are much less effective photon scatterers. Similarly, the relevant mass for electrons bound in atoms is the mass of the entire atom, and hence bound electrons pose a very small Thomson scattering cross section.

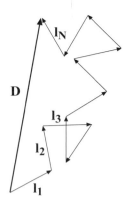

Figure 3.4 The net advance of a photon performing a random walk consisting of N steps, each decribed by a vector l_i, is the vector $\mathbf{D}$.

The expectation value of the term in parentheses in Eq. 3.43 is zero because it is a sum over many vector dot products, each with a random angle, and hence with both positive and negative cosines contributing equally. The linear distance covered in a random walk is therefore

$$\langle \mathbf{D}^2 \rangle^{1/2} = D = \sqrt{N}\, l. \tag{3.45}$$

To gain some intuition, it is instructive to calculate how long it takes a photon to travel from the center of the Sun, where most of the energy is produced, to the surface.[4] From Eq. 3.45, traveling a distance $r_\odot$ will require $N = r_\odot^2/l^2$ steps in the random walk. Each step requires a time l/c. Thus, the total time for the photon to emerge from the Sun is

$$\tau_{\text{rw}} \approx \frac{l}{c}\frac{r_\odot^2}{l^2} = \frac{r_\odot^2}{lc} = \frac{(7 \times 10^{10}\ \text{cm})^2}{10^{-1}\ \text{cm} \times 3 \times 10^{10}\ \text{cm s}^{-1}}$$

$$= 1.6 \times 10^{12}\ \text{s} = 52,000\ \text{yr}. \tag{3.46}$$

Thus, if the nuclear reactions powering the Sun were to suddenly switch off, we would not notice[5] anything unusual for 50,000 years.

With this background, we can now derive the equation that relates the temperature profile, $T(r)$, to the flow of radiative energy through a star. The small mean free path of photons inside the Sun and the very numerous scatterings, absorptions, and reemissions every photon undergoes reaffirms that locally, every volume element inside the Sun radiates as a blackbody to a very good approximation. However, there is a net flow of radiation energy outwards, meaning there is some small anisotropy (a preferred direction), and

---

[4] In reality, of course, it is not the same photon that travels this path. In every interaction, the photon can transfer energy to the particle it scatters on, or distribute its original energy among several photons that emerge from the interaction. Hence, the photon energy is strongly "degraded" during the passage through the Sun.

[5] In fact, even then, nothing dramatic would happen. As we will see in section 3.9, a slow contraction of the Sun would begin, with a timescale of $\sim 10^7$ yr. Over $\sim 10^5$ yr, the solar radius would only shrink by $\sim 1\%$, which is small but discernible.

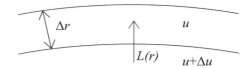

Figure 3.5 Radiative diffusion of energy between volume shells in a star, driven by the gradient in the thermal energy density.

implying there is a higher energy density, at smaller radii than at larger radii. The net flow of radiation energy through a mass shell at radius r, per unit time, is just $L(r)$ (see Fig. 3.5). This must equal the excess energy in the shell, compared to a shell at larger radius, divided by the time it takes this excess energy to flow across the shell's width Δr. The excess energy density is Δu, which multiplied by the shell's volume gives the total excess energy of the shell, $4\pi r^2 \Delta r \Delta u$. The time for the photons to cross the shell in a random walk is $(\Delta r)^2/lc$. Thus,

$$L(r) \approx -\frac{4\pi r^2 \Delta r \Delta u}{(\Delta r)^2/lc} = -4\pi r^2 lc \frac{\Delta u}{\Delta r}. \tag{3.47}$$

A more rigorous derivation of this equation adds a factor $1/3$ on the right-hand side, which comes about from an integration of $\cos^2\theta$ over a solid angle (see the derivation of the equation of radiation pressure, Eq. 3.74, below). Including this factor and replacing the differences with differentials, we obtain

$$\frac{L(r)}{4\pi r^2} = -\frac{cl}{3}\frac{du}{dr}. \tag{3.48}$$

Note that this is, in effect, a **diffusion equation**, describing the outward flow of energy. The left-hand side is the energy flux. On the right-hand side, du/dr is the gradient in energy density, and $-cl/3$ is a diffusion coefficient that sets the proportionality relating the energy flow and the energy density gradient. The opacity, as reflected in the mean free path l, controls the flow of radiation through the star. For low opacity (large l), the flow will be relatively unobstructed, and hence the luminosity will be high, and vice versa.

Since at every radius the energy density is close to that of blackbody radiation, then (Eq. 2.9)

$$u = aT^4, \tag{3.49}$$

and

$$\frac{du}{dr} = \frac{du}{dT}\frac{dT}{dr} = 4aT^3\frac{dT}{dr}. \tag{3.50}$$

Substituting in Eq. 3.48, and expressing l as $(\kappa\rho)^{-1}$, we obtain the **equation of radiative energy transport**,

$$\frac{dT(r)}{dr} = -\frac{3L(r)\kappa(r)\rho(r)}{4\pi r^2 4acT^3(r)}. \tag{3.51}$$

From Eqs. 3.48 and 3.49, together with an estimate of the mean free path, we can make an order-of-magnitude prediction of the Sun's luminosity. Approximating $-du/dr$ with $\sim u/r_\odot = aT^4/r_\odot$, we have

$$L_\odot \sim 4\pi r_\odot^2 \frac{cl}{3} \frac{aT^4}{r_\odot}. \tag{3.52}$$

Based on the virial theorem and the Sun's mass and radius, we obtained in Eq. 3.33 an estimate of the Sun's internal temperature, $T_{\text{vir}} \sim 4 \times 10^6$ K. Using this as a typical temperature and taking $l = 0.1$ cm, we find

$$L_\odot \sim \frac{4\pi}{3} 7 \times 10^{10} \text{ cm} \times 3 \times 10^{10} \text{ cm s}^{-1} \times 10^{-1} \times 7.6 \times 10^{-15} \text{ cgs} \times (4 \times 10^6 \text{ K})^4$$
$$= 2 \times 10^{33} \text{ ergs}^{-1}, \tag{3.53}$$

in reasonable agreement with the observed solar luminosity, $L_\odot = 3.8 \times 10^{33}$ ergs^{-1}. (We have abbreviated above the units of the radiation constant, a, as cgs.) The above estimate can also be used to argue that, based on its observed luminosity, the Sun must be composed primarily of ionized hydrogen. If the Sun were composed of, say, ionized carbon, the mean particle mass would be $\bar{m} \approx 12m_H/7 \approx 2m_H$, rather than $m_H/2$. Equation 3.33 would then give a virial temperature that is 4 times as high, resulting in a luminosity prediction in Eq. 3.52 that is too high by two orders of magnitude.

3.4 Energy Conservation

We will see that the luminosity of a star is produced by nuclear reactions, with output energies that depend on the local conditions (density and temperature) and hence on r. Let us define $\epsilon(r)$ as the power produced per unit mass of stellar material. Energy conservation means that the addition to a star's luminosity due to the energy production in a thin shell at radius r is

$$dL = \epsilon \, dm = \epsilon \rho 4\pi r^2 dr, \tag{3.54}$$

or

$$\frac{dL(r)}{dr} = 4\pi r^2 \rho(r)\epsilon(r), \tag{3.55}$$

which is the **equation of energy conservation**.

3.5 The Equations of Stellar Structure

We have derived four coupled first-order differential equations describing stellar structure. Let us rewrite them here:

$$\frac{dP(r)}{dr} = -\frac{GM(r)\rho(r)}{r^2}, \tag{3.56}$$

$$\frac{dM(r)}{dr} = 4\pi r^2 \rho(r), \tag{3.57}$$

$$\frac{dT(r)}{dr} = -\frac{3L(r)\kappa(r)\rho(r)}{4\pi r^2 4ac T(r)^3}, \tag{3.58}$$

$$\frac{dL(r)}{dr} = 4\pi r^2 \rho(r)\epsilon(r). \tag{3.59}$$

We can define four boundary conditions for these equations, for example,

$$M(r = 0) = 0, \tag{3.60}$$
$$L(r = 0) = 0, \tag{3.61}$$
$$P(r = r_*) = 0, \tag{3.62}$$
$$M(r = r_*) = M_*, \tag{3.63}$$

where M_* is the total mass of the star. (In reality, at the radius r_* of the photosphere of the star, P does not really go completely to zero, nor do T and ρ, and more sophisticated boundary conditions are required, which account for the processes in the photosphere.)

To these four differential equations we need to add three equations connecting the pressure, the opacity, and the energy production rate of the gas with its density, temperature, and composition:

$$P = P(\rho, T, \text{composition}), \tag{3.64}$$
$$\kappa = \kappa(\rho, T, \text{composition}), \tag{3.65}$$
$$\epsilon = \epsilon(\rho, T, \text{composition}). \tag{3.66}$$

$P(\rho, T)$ is usually called **the equation of state.** Each of these three functions will depend on the composition through the **element abundances** and the ionization states of each element in the gas. It is common in astronomy to parametrize the mass abundances of hydrogen, helium, and the heavier elements (the latter are often referred to collectively by the term "metals") as

$$X \equiv \frac{\rho_H}{\rho}, \quad Y \equiv \frac{\rho_{He}}{\rho}, \quad Z \equiv \frac{\rho_{metals}}{\rho}. \tag{3.67}$$

We have thus ended up with seven coupled equations defining the seven unknown functions: $P(r)$, $M(r)$, $\rho(r)$, $T(r)$, $\kappa(r)$, $L(r)$, and $\epsilon(r)$. As there are four boundary conditions for the four first-order differential equations, if there is a solution, it is unique. This is usually expressed in the form of the *Vogt-Russell conjecture*, which states that the properties and evolution of an isolated star are fully determined by its initial mass and its chemical abundances. These determine the star's observable parameters: its surface temperature, radius, and luminosity. Two variables that we have neglected in this treatment, and that have minor influence on stellar structure, are stellar rotation and magnetic fields. To proceed, we need to define the three functions, P, κ, and ϵ.

3.6 The Equation of State

Different equations of state $P(\rho, T, X, Y, Z)$ apply for different ranges of gas density, temperature, and abundance. Under the conditions in most normal stars, the equation of state of a classical, nonrelativistic, ideal gas, provides a good description. Consider, for example, such a gas, composed of three different kinds of particles, each with its own mass m_i and density n_i. The mean particle mass will be

$$\bar{m} = \frac{n_1 m_1 + n_2 m_2 + n_3 m_3}{n_1 + n_2 + n_3} = \frac{\rho}{n}. \tag{3.68}$$

The gas pressure will then be

$$P_g = nkT = \frac{\rho}{\bar{m}} kT. \tag{3.69}$$

The mean mass will depend on the chemical abundance and ionization state of the gas. As we have already seen, for completely ionized pure hydrogen,

$$\bar{m} = \frac{m_H}{2}, \tag{3.70}$$

and therefore $\bar{m}/m_H = 0.5$.

More generally, the number densities of hydrogen, helium, or an element of atomic mass number A (i.e., an element with a total of A protons and neutrons in each atomic nucleus) will be

$$n_H = \frac{X\rho}{m_H}, \quad n_{\mathrm{He}} = \frac{Y\rho}{4m_H}, \quad n_A = \frac{Z_A\rho}{Am_H}, \tag{3.71}$$

where Z_A is the mass abundance of an element of atomic mass number A. Complete ionization of hydrogen results in two particles (an electron and a proton); of helium, three particles (two electrons and a nucleus); and of an atom with atomic number $\mathcal{Z}$ (i.e., with $\mathcal{Z}$ protons or electrons), $\mathcal{Z} + 1$ particles, which for heavy enough atoms is always close to $A/2$. Thus, for an ionized gas we will have

$$n = 2n_H + 3n_{\mathrm{He}} + \sum \frac{A}{2} n_A = \frac{\rho}{m_H} \left(2X + \frac{3}{4}Y + \frac{1}{2}Z \right)$$

$$= \frac{\rho}{2m_H} \left(3X + \frac{Y}{2} + 1 \right), \tag{3.72}$$

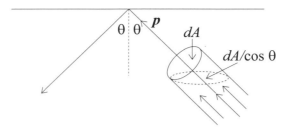

Figure 3.6 Calculation of radiation pressure. A beam of photons with blackbody intensity B strikes the wall of a container at an angle θ to its perpendicular. The projected area of the beam, dA, is increased by $1/\cos\theta$, and therefore the power reaching the wall per unit area is decreased to $B\cos\theta$. Since a photon's momentum, p, is its energy divided by c, the momentum flux in the beam is B/c. Every reflection of a photon transfers twice its perpendicular component of momentum, and therefore the momentum transfer per unit time and per unit area, i.e., the pressure, is $2B\cos^2\theta/c$. The total pressure is obtained by integrating over all angles of the beams that approach the wall.

where we have used the fact that $X + Y + Z = 1$. Thus,

$$\frac{\bar{m}}{m_H} = \frac{\rho}{n m_H} = \frac{2}{1 + 3X + 0.5Y} \tag{3.73}$$

for a totally ionized gas. For solar abundances, $X = 0.71$, $Y = 0.27$, $Z = 0.02$, and therefore $\bar{m}/m_H = 0.61$. In the central regions of the Sun, about half of the hydrogen has already been converted into helium by nuclear reactions, and as a result $X = 0.34$, $Y = 0.64$, and $Z = 0.02$, giving $\bar{m}/m_H = 0.85$.

In addition to the kinetic gas pressure, the photons in a star exert **radiation pressure**. Let us digress briefly, and derive the equation of state for this kind of pressure. Consider photons inside a blackbody radiator with an intensity given by the Planck function, $I_\nu = B_\nu$, which, when integrated over wavelength, we denoted as B. As illustrated in Fig. 3.6, the energy arriving at the surface of the radiator per unit time, per unit area, at some angle θ to the perpendicular to the surface, is $B\cos\theta$, because the area of the beam, when projected onto the wall of the radiator, is increased by $1/\cos\theta$. Now, consider the photons in the beam, which strike the fully reflective surface of the radiator at the angle θ. Every photon of energy E has momentum $p = E/c$. When reflected, it transmits to the surface a momentum $\Delta p = (2E/c)\cos\theta$. Therefore, there is a second factor of $\cos\theta$ that must be applied to the incoming beam. The rate of momentum transfer per unit area will be obtained by integrating over all angles at which the photons hit the surface. But, the rate of momentum transfer (i.e., the force) per unit area is, by definition, the pressure. Thus,

$$P = \frac{F}{A} = \frac{dp/dt}{A} = \frac{2}{c}\int_{\pi/2}^{\pi} B\cos^2\theta\sin\theta\,d\theta\,d\phi = \frac{4\pi}{3c}B = \frac{1}{3}u, \tag{3.74}$$

where in the last equality we have used the previously found relation (Eq. 2.4) between intensity and energy density. Note that the derivation above applies not only to photons,

Figure 3.7 *Left:* A free electron accelerating in the Coulomb potential of an ion emits bremsstrahlung, or "free–free" radiation. *Right:* In the inverse process of free–free absorption, a photon is absorbed by a free electron. The process is possible only if a neighboring ion, which can share some of the photon's momentum, is present.

but to any particles with an isotropic velocity distribution, and with kinetic energies large compared to their rest-mass energies, so that the relation $p \approx E/c$ (which is exact for photons) is a good approximation. Thus, the equation of state in which the radiation pressure equals one-third of the thermal energy density holds for any ultrarelativistic gas.

Returning to the case of the pressure due to a thermal photon gas inside a star, we can write

$$P_{\mathrm{rad}} = \tfrac{1}{3}u = \tfrac{1}{3}aT^4, \tag{3.75}$$

which under some circumstances can become important or dominant (see Problems 2 and 3). The full equation of state for normal stars will therefore be

$$P = P_g + P_{\mathrm{rad}} = \frac{\rho kT}{\bar{m}} + \frac{1}{3}aT^4. \tag{3.76}$$

We will see in chapter 4 that the conditions in white dwarfs and in neutron stars dictate equations of state that are very different from this form.

3.7 Opacity

Like the equation of state, the opacity, κ, at every radius in the star will depend on the density, the temperature, and the chemical composition at that radius. We have already mentioned one important source of opacity, Thomson scattering of photons off free electrons. Let us now calculate correctly the electron density for an ionized gas of arbitrary abundance (rather than pure hydrogen, as before):

$$n_e = n_H + 2n_{\mathrm{He}} + \sum \frac{A}{2} n_A = \frac{\rho}{m_H}\left(X + \frac{2}{4}Y + \frac{1}{2}Z\right) = \frac{\rho}{2m_H}(1+X), \tag{3.77}$$

where we have again assumed that the number of electrons in an atom of mass number A is $A/2$. Therefore,

$$\kappa_{\mathrm{es}} = \frac{n_e\sigma_T}{\rho} = \frac{\sigma_T}{2m_H}(1+X) = (1+X)\,0.2\ \mathrm{cm}^2\ \mathrm{g}^{-1}. \tag{3.78}$$

In regions of a star with relatively low temperatures, such that some or all of the electrons are still bound to their atoms, three additional processes that are important sources of opacity are **bound–bound**, **bound–free** (also called **photoionization**), and **free–free** absorption. In bound–bound and bound–free absorption, which we have already discussed in the context of photospheric absorption features, an atom or ion is excited to a higher energy level, or ionized to a higher degree of ionization, by absorbing a photon. Free–free absorption is the inverse process of free–free emission, often called **bremsstrahlung** ("braking radiation" in German). In free–free emission (see Fig. 3.7), a free electron is accelerated by the electric potential of an ion, and as a result radiates. Thus, in free–free absorption, a photon is absorbed by a free electron and an ion, which share the photon's momentum and energy. All three processes depend on photon wavelength, in addition to gas temperature, density, and composition.

When averaged over all wavelengths, the *mean opacity* due to both bound–free absorption and free–free absorption behave approximately as

$$\bar{\kappa}_{\mathrm{bf,ff}} \sim \frac{\rho}{T^{3.5}}, \tag{3.79}$$

which is called a Kramers opacity law. This behavior holds only over limited ranges in temperature and density. For example, free–free absorption actually *increases* with temperature at low temperature and density, with the increase in free electron density. Similarly, bound–free opacity cuts off at high temperatures at which the atoms are fully ionized. Additional sources of opacity, significant especially in low-mass stars, are molecules and H^- ions.[6]

3.8 Scaling Relations on the Main Sequence

From the equations we have derived so far, we can already deduce and understand the observed functional forms of the mass–luminosity relation, $L \sim M^{\alpha}$, and the effective-temperature–luminosity relation, $L \sim T_E^8$, that are observed for main sequence stars. Let us assume, for simplicity, that the functions $P(r)$, $M(r)$, $\rho(r)$, and $T(r)$ are roughly power laws, i.e., $P(r) \sim r^{\beta}$, $M(r) \sim r^{\gamma}$, etc. If so, we can immediately write the first three differential equations (Eqs. 3.56, 3.57, and 3.58) as scaling relations,

$$P \sim \frac{M\rho}{r}, \tag{3.80}$$

$$M \sim r^3 \rho, \tag{3.81}$$

and

$$L \sim \frac{T^4 r}{\kappa \rho} \tag{3.82}$$

---

[6] The negative H^- ion forms when a second electron attaches (with a quite weak bond) to a hydrogen atom.

(just as, instead of solving a differential equation, say, $df/dx = x^4$, we can write directly $f \sim x^5$). For moderately massive stars, the pressure will be dominated by the kinetic gas pressure, and the opacity by electron scattering. Therefore,

$$P \sim \rho T \tag{3.83}$$

and (Eq. 3.78)

$$\kappa = \text{const.} \tag{3.84}$$

Equating 3.80 and 3.83, we find

$$T \sim \frac{M}{r} \tag{3.85}$$

(which is basically just the virial theorem again, for a nonrelativistic, classical, ideal gas— Eq. 3.27). Substituting this into 3.82, and using 3.81 to express $r^3 \rho$, we find

$$L \sim M^3, \tag{3.86}$$

as observed for main-sequence stars more massive than the Sun.

Equation 3.85 also suggests that $r \sim M$ on the main sequence. To see this, consider a star that is forming from a mass M that is contracting under its own gravity and heating up (star formation is discussed in some detail in chapter 5). The contraction will stop, and an equilibrium will be set up, once the density and the temperature in the core are high enough for the onset of nuclear reactions. We will see that the nuclear power density depends mainly on temperature. Thus, for any initial mass, r will stop shrinking when a particular core temperature is reached. Therefore, the internal temperature T is comparable in all main-sequence stars (i.e, it is weakly dependent on mass, and hence approximately constant), and

$$r \sim M. \tag{3.87}$$

Detailed models confirm that the core temperature varies only by a factor ≈ 4 over a range of ~ 100 in mass on the main sequence. With $r \sim M$, we see from Eq. 3.81 that the density of a star decreases as M^{-2}, so that more massive stars will have low density, and low-mass stars will have high density.

Proceeding to low-mass stars, the high density means there is a dominant role for bound–free and free–free opacity,

$$\kappa \sim \frac{\rho}{T^{3.5}}. \tag{3.88}$$

Since $T \sim$ const., $r \sim M$, and $\rho \sim M^{-2}$, then $\kappa \sim \rho \sim M^{-2}$, and Eq. 3.82 gives

$$L \sim \frac{T^4 r}{\kappa \rho} \sim \frac{r}{\rho^2} \sim M^5, \tag{3.89}$$

as seen in low-mass stars.

For the most massive stars, the low gas density will make radiation pressure dominant in the equation of state (see Problem 3),

$$P \sim T^4, \tag{3.90}$$

and electron scattering, with $\kappa = \mathrm{const.}$, will again be the main source of opacity. Equating with 3.80 and substituting for T^4 in Eq. 3.82, we find

$$L \sim M, \tag{3.91}$$

a flattening of the mass–luminosity relationship that is, in fact, observed for the most massive stars.

Finally, we can also reproduce the functional dependence of the main sequence in the H-R diagram. We saw that $L \sim M^5$ for low-mass stars and $L \sim M^3$ for moderately massive stars. Let us then take an intermediate slope, $L \sim M^4$, as representative. Since $r \sim M$, then

$$\sigma T_E^4 = \frac{L}{4\pi r_*^2} \sim \frac{M^4}{M^2} \sim M^2 \sim L^{1/2}, \tag{3.92}$$

so

$$L \sim T_E^8, \tag{3.93}$$

as observed.

We have thus seen that the mass vs. luminosity relation and the surface-temperature vs. luminosity relation of main-sequence stars are simply consequences of the different sources of pressure and opacity in stars of different masses, and of the fact that the onset of nuclear hydrogen burning keeps the core temperatures of all main-sequence stars in a narrow range. The latter fact is elucidated below.

3.9 Nuclear Energy Production

The last function we still need to describe is the power density $\epsilon(\rho, T, X, Y, Z)$. To see that the energy source behind ϵ must be nuclear burning, we consider the alternatives. Suppose that the source of the Sun's energy were gravitational, i.e., that the Sun had radiated until now the potential energy liberated by contracting from infinity to its present radius. From the virial theorem, we saw that the thermal energy resulting from such a contraction is minus one-half the gravitational energy,

$$E_{\mathrm{gr}} = -2E_{\mathrm{th}}. \tag{3.94}$$

Therefore, the other half of the gravitational energy released by the contraction, and which the Sun could have radiated, is

$$E_{\mathrm{rad}} \sim \frac{1}{2}\frac{GM_\odot^2}{r_\odot}. \tag{3.95}$$

To see how long the Sun could have shined at its present luminosity with this energy source, we divide this energy by the solar luminosity. This gives the so-called **Kelvin-Helmholtz timescale**,

$$\tau_{\rm kh} \sim \frac{1}{2} \frac{GM_\odot^2}{r_\odot} \frac{1}{L_\odot} = \frac{6.7 \times 10^{-8} \text{ cgs} \times (2 \times 10^{33} \text{ g})^2}{2 \times 7 \times 10^{10} \text{ cm} \times 3.8 \times 10^{33} \text{ erg s}^{-1}}$$

$$= 5 \times 10^{14} \text{ s} = 1.6 \times 10^7 \text{ yr}. \tag{3.96}$$

The geological record shows that the Earth and Moon have existed for over 4 billion years, and that the Sun has been shining with about the same luminosity during all of this period. A similar calculation shows that chemical reactions (e.g., if the Sun were producing energy by combining hydrogen and oxygen into water) are also not viable for producing the solar luminosity for so long.

A viable energy source for the Sun and other main-sequence stars is nuclear fusion of hydrogen into helium. Most of the nuclear energy of the Sun comes from a chain of reactions called the **p-p chain**. The first step is the reaction

$$p + p \to d + e^+ + \nu_e, \tag{3.97}$$

where d designates a deuteron, composed of a proton and a neutron. As we will see in section 3.10, the timescale for this process inside the Sun is 10^{10} yr. The timescale is so long mainly because the reaction proceeds via the weak interaction (as is evidenced by the emission of a neutrino). The positron, the deuteron, and the neutrino share an energy of 0.425 MeV. Once the reaction occurs, the positron quickly annihilates with an electron, producing two 0.511-MeV γ-ray photons. The neutrino, having a weak interaction with matter, escapes the Sun and carries off its energy, which has a mean of 0.26 MeV. The remaining kinetic energy and photons quickly thermalize by means of frequent matter–matter and matter–photon collisions. Typically, within 1 s, the deuteron will merge with another proton to form ^{3}He:

$$p + d \to {}^3\text{He} + \gamma, \tag{3.98}$$

with a total energy release (kinetic + the γ-ray photon) of 5.49 MeV. Finally, on a timescale of 300,000 years, we have

$$^3\text{He} + {}^3\text{He} \to {}^4\text{He} + p + p, \tag{3.99}$$

with a kinetic energy release of 12.86 MeV. Every time this three-step chain occurs twice, four protons are converted into a ^{4}He nucleus, two neutrinos, photons, and kinetic energy. The total energy released per ^{4}He nucleus is thus

$$(4 \times 0.511 + 2 \times 0.425 + 2 \times 5.49 + 12.86) \text{ MeV} = 26.73 \text{ MeV}. \tag{3.100}$$

Deducting the 2×0.511 MeV from the annihilation of two preexisting electrons, we find that this is just the rest-mass difference between four free protons and a ^{4}He nucleus:

$$[m(4p) - m({}^4\text{He})]c^2 = 25.71 \text{ MeV} = 0.7\% \, m(4p)c^2. \tag{3.101}$$

Thus, the rest-mass-to-energy conversion efficiency of the p-p chain is 0.7%. The time for the Sun to radiate away just 10% of the energy available from this source is

$$\begin{aligned}
\tau_{\text{nuc}} &= \frac{0.1 \times 0.007 \times M_\odot c^2}{L_\odot} \\
&= \frac{0.1 \times 0.007 \times 2 \times 10^{33} \text{ g} \times (3 \times 10^{10} \text{ cm s}^{-1})^2}{3.8 \times 10^{33} \text{ erg s}^{-1}} \\
&= 3.3 \times 10^{17} \text{ s} = 10^{10} \text{ yr.}
\end{aligned} \tag{3.102}$$

In other words, in terms of energy budget, hydrogen fusion can easily produce the solar luminosity over the age of the Solar System.

Next, we need to see if the conditions in the Sun are suitable for these reactions to actually take place. Consider two nuclei with atomic numbers (i.e., number of protons per nucleus) $\mathcal{Z}_A$ and $\mathcal{Z}_B$. The strong interaction produces a short-range attractive force between the nuclei on scales smaller than

$$r_0 \approx 1.4 \times 10^{-13} \text{ cm.} \tag{3.103}$$

The strong interaction goes to zero at larger distances, and the Coulomb repulsion between the nuclei takes over. The Coulomb energy barrier is

$$E_{\text{coul}} = \frac{\mathcal{Z}_A \mathcal{Z}_B e^2}{r}, \tag{3.104}$$

which at r_0 is of order

$$E_{\text{coul}}(r_0) \approx \mathcal{Z}_A \mathcal{Z}_B \text{ MeV.} \tag{3.105}$$

Figure 3.8 shows schematically the combined nuclear (strong) and electrostatic (Coulomb) potential. In the reference frame of one of the nuclei, the other nucleus, with kinetic energy E, can classically approach only to a distance

$$r_1 = \frac{\mathcal{Z}_A \mathcal{Z}_B e^2}{E}, \tag{3.106}$$

where it will be repelled away. At a typical internal stellar temperature of 10^7 K, the kinetic energy of a nucleus is $1.5kT \sim 1$ keV. The characteristic kinetic energy is thus of order 10^{-3} of the energy required to overcome the Coulomb barrier. Typical nuclei will approach each other only to a separation $r_1 \sim 10^{-10}$ cm, 1000 times larger than the distance at which the strong nuclear binding force operates. Perhaps those nuclei that are in the high-energy tail of the Maxwell-Boltzmann distribution can overcome the barrier? The fraction of nuclei with such energies is

$$e^{-E/kT} \approx e^{-1000} \approx 10^{-434}. \tag{3.107}$$

The number of protons in the Sun is

$$N_p \approx \frac{M_\odot}{m_H} = \frac{2 \times 10^{33} \text{ g}}{1.7 \times 10^{-24} \text{ g}} \approx 10^{57}. \tag{3.108}$$

Thus, there is not a single nucleus in the Sun (or, for that matter, in all the stars in the observable Universe) with the kinetic energy required classically to overcome the Coulomb barrier and undergo nuclear fusion with another nucleus.

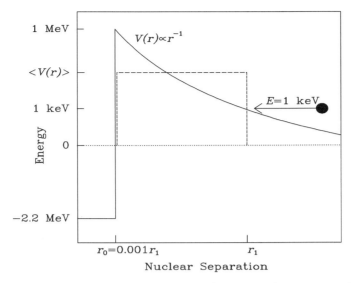

Figure 3.8 Schematic illustration of the potential energy $V(r)$ between two nuclei as a function of separation r. For two protons, the Coulomb repulsion reaches a maximum, with $V(r) \sim 1$ MeV at $r = r_0$, at which point the short-range nuclear force sets in and binds the nuclei (negative potential energy). Two nuclei with relative kinetic energy of ~ 1 keV, typical for the temperatures in stellar interiors, can classically approach each other only to within a separation r_1, 1000 times greater than r_0. The dashed rectangle is a *rectangular barrier* of height $\langle V(r) \rangle \approx 3E/2$, which we use to approximate the Coulomb barrier in our calculation of the probability for quantum tunneling through the potential.

Fortunately, **quantum tunneling** through the barrier allows nuclear reactions to take place after all. To see this, let us describe this two-body problem by means of the time-independent Schrödinger equation, for a wave function Ψ in a spherically symmetric potential $V(r)$:

$$\frac{\hbar^2}{2\mu} \nabla^2 \Psi = [V(r) - E]\Psi, \tag{3.109}$$

where the reduced mass of the two nuclei, of masses m_A and m_B, is

$$\mu \equiv \frac{m_A m_B}{m_A + m_B}. \tag{3.110}$$

In our case, the potential is

$$V(r) = \frac{\mathcal{Z}_A \mathcal{Z}_B e^2}{r}, \tag{3.111}$$

and E is the kinetic energy. Let us obtain an order-of-magnitude solution to the Schrödinger equation. By our definition of r_1, the radius of closest classical approach, we have $V(r_1) = E$.

We can then write $V(r) = Er_1/r$, and the mean, volume-averaged, height of the potential between r_1 and $r_0 \ll r_1$ is

$$\langle V(r) \rangle = \frac{\int_{r_0}^{r_1} 4\pi r^2 V(r) dr}{\int_{r_0}^{r_1} 4\pi r^2 dr} \approx \frac{3}{2} E. \tag{3.112}$$

Approximating $V(r)$ with a constant function of this height (a "rectangular barrier"), the radial component of the Schrödinger equation becomes

$$\frac{\hbar^2}{2\mu} \frac{1}{r} \frac{d^2(r\Psi)}{dr^2} \approx \frac{E}{2} \Psi, \tag{3.113}$$

which has a solution

$$\Psi = A \frac{e^{\beta r}}{r}, \quad \beta = \frac{\sqrt{\mu E}}{\hbar}. \tag{3.114}$$

(The second independent solution of the equation, with an amplitude that rises with decreasing radius, is unphysical.) The wave function amplitude, squared, is proportional to the probability density for a particle to be at a given location. Multiplying the ratio of the probability densities by the ratio of the volume elements, $4\pi r_0^2 dr$ and $4\pi r_1^2 dr$, thus gives the probability that a nucleus will tunnel from r_1 to within $r_0 \ll r_1$ of the other nucleus:

$$\frac{|\Psi(r_0)|^2 \, r_0^2}{|\Psi(r_1)|^2 \, r_1^2} = \frac{e^{2\beta r_0}}{e^{2\beta r_1}} \approx e^{-2\beta r_1} = \exp\left(-\frac{2\sqrt{\mu E}}{\hbar} \frac{\mathcal{Z}_A \mathcal{Z}_B e^2}{E}\right)$$

$$= \exp\left(-\frac{2\sqrt{\mu}}{\hbar} \mathcal{Z}_A \mathcal{Z}_B e^2 \frac{1}{\sqrt{E}}\right). \tag{3.115}$$

A full solution of the Schrödinger equation gives the same answer, but with an additional factor $\pi/\sqrt{2}$ in the exponential. If we recall the definition of the fine-structure constant,

$$\alpha = \frac{e^2}{\hbar c} \approx \frac{1}{137}, \tag{3.116}$$

and define an energy

$$E_G = (\pi \alpha \mathcal{Z}_A \mathcal{Z}_B)^2 2\mu c^2, \tag{3.117}$$

then the probability of penetrating the Coulomb barrier simplifies to the function

$$g(E) = e^{-\sqrt{E_G/E}}. \tag{3.118}$$

E_G is called the **Gamow energy** and $g(E)$ is called the **Gamow factor**. For two protons,

$$E_G = \left(\pi \frac{1}{137} \times 1 \times 1\right)^2 2\frac{1}{2} m_p c^2 \approx 500 \text{ keV}. \tag{3.119}$$

(It is convenient to remember that the rest energy of a proton, $m_p c^2$, is 0.94 GeV.) Thus, for the typical kinetic energy of particles in the Sun's core, $E \sim 1$ keV, we find

$g(E) \sim e^{-22} \sim 10^{-10}$. While this probability, for a given pair or protons, is still small, it is considerably larger than the classical probability we found in Eq. 3.107.

3.10 Nuclear Reaction Rates

Even if tunneling occurs, and two nuclei are within the strong force's interaction range, the probability of a nuclear reaction will still depend on a nuclear cross section, which will generally depend inversely on the kinetic energy. Thus, the total cross section for a nuclear reaction involving ingredient nuclei A and B is

$$\sigma_{AB}(E) = \frac{S_0}{E} e^{-\sqrt{E_G/E}}, \tag{3.120}$$

where S_0 is a constant, or a weak function of energy, with units of [area]$\times$[energy]. S_0 for a given nuclear reaction is generally derived from accelerator experiments, or is calculated theoretically.

The number of reactions per nucleus A as it traverses a distance dx in a field with a density n_B of "target" nuclei B is

$$dN_A = n_B \sigma_{AB} dx. \tag{3.121}$$

If we divide both sides by dt, then the number of reactions per nucleus A per unit time is

$$\frac{dN_A}{dt} = n_B \sigma_{AB} v_{AB}, \tag{3.122}$$

where v_{AB} is the relative velocity between the nuclei.

From Eq. 3.122, we can proceed to find the power density function, $\epsilon(\rho, T, X, Y, Z)$, needed to solve the equations of stellar structure. Multiplying by the density n_A of nuclei A will give the number of reactions per unit time and per unit volume, i.e., the reaction rate per unit volume,

$$\mathcal{R}_{AB} = n_A n_B \sigma_{AB} v_{AB}. \tag{3.123}$$

If every reaction releases an energy Q, multiplying by Q gives the power per unit volume. Dividing by ρ then gives the power per unit mass, rather than per unit volume:

$$\epsilon = n_A n_B \sigma_{AB} v_{AB} Q / \rho. \tag{3.124}$$

Recalling that

$$n_A = \frac{\rho X_A}{A_A m_H}, \quad n_B = \frac{\rho X_B}{A_B m_H}, \tag{3.125}$$

with X_A and X_B symbolizing the mass abundances, and A_A and A_B the atomic mass numbers of the two nucleus types, ϵ can be expressed as

$$\epsilon = \frac{\rho X_A X_B}{m_H^2 A_A A_B} \sigma_{AB} v_{AB} Q. \tag{3.126}$$

In reality, the nuclei in a gas will have a distribution of velocities, so every velocity has some probability of occurring. Hence, ϵ can be obtained by averaging $v_{AB}\sigma_{AB}$ over all velocities, with each velocity weighted by its probability, $P(v_{AB})$:

$$\epsilon = \frac{\rho X_A X_B}{m_H^2 A_A A_B} \langle \sigma_{AB} v_{AB} \rangle Q,$$ (3.127)

with

$$\langle \sigma_{AB} v_{AB} \rangle = \int_0^\infty \sigma_{AB} v_{AB} P(v_{AB}) dv_{AB}.$$ (3.128)

A classical nonrelativistic gas will have a distribution of velocities described by the Maxwell-Boltzmann distribution. The *relative* velocities of nuclei A and B will also follow a Maxwell-Boltzmann distribution,

$$P(v)dv = 4\pi \left(\frac{\mu}{2\pi kT} \right)^{3/2} v^2 \exp \left(-\frac{\mu v^2}{2kT} \right) dv,$$ (3.129)

but with a mass represented by the reduced mass of the particles, μ. For brevity, we have omitted here the subscript AB from the velocities.

Inserting 3.120 and 3.129 into 3.128, and changing the integration variable from velocity to kinetic energy using $E = \frac{1}{2}\mu v^2$, $dE = \mu v dv$, we obtain

$$\langle \sigma v \rangle = \left(\frac{8}{\pi \mu} \right)^{1/2} \frac{S_0}{(kT)^{3/2}} \int_0^\infty e^{-E/kT} e^{-\sqrt{E_G/E}} dE.$$ (3.130)

The integrand in this expression,

$$f(E) = e^{-E/kT} e^{-\sqrt{E_G/E}},$$ (3.131)

is composed of the product of two exponential functions, one (from the Boltzmann distribution) falling with energy, and the other (due to the Gamow factor embodying the Coulomb repulsion) rising with energy. Obviously, $f(E)$ will have a narrow maximum at some energy E_0, at which most of the reactions take place (see Fig. 3.9). The maximum of $f(E)$ is easily found by taking its derivative and equating to zero. It is at

$$E_0 = \left(\frac{kT}{2} \right)^{2/3} E_G^{1/3}.$$ (3.132)

A Taylor expansion of $f(E)$ around E_0 shows $f(E)$ can be approximated by a Gaussian with a width parameter (i.e., the "σ", or standard deviation, of the Gaussian $e^{-x^2/2\sigma^2}$) of

$$\Delta = \frac{2^{1/6}}{3^{1/2}} E_G^{1/6} (kT)^{5/6}.$$ (3.133)

The value of the integral can therefore be approximated well with the area of the Gaussian, $\sqrt{2\pi} f(E_0)\Delta$ (see Problem 7). Replacing in 3.127, we obtain the final expression for the power density due to a given nuclear reaction,

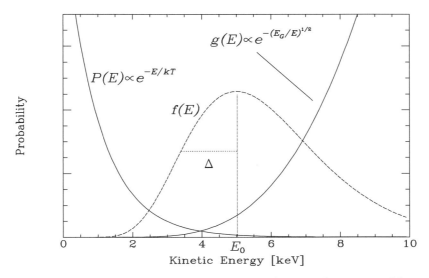

Figure 3.9 The Boltzmann probability distribution, $P(E)$, the Gamow factor, $g(E)$, and their product, $f(E)$. $P(E)$ is shown for $kT = 1$ keV, $g(E)$ is for the case of two protons, with $E_G = 500$ keV, and both $g(E)$ and $f(E)$ have been scaled up by large factors for display purposes. The rarity of protons with large kinetic energies, as described by $P(E)$, combined with the Coulomb barrier, embodied by $g(E)$, limit the protons taking part in nuclear reactions to those with energies near E_0, where $f(E)$ peaks. $f(E)$ can be approximated as a Gaussian centered at E_0 ($E_0 \approx 5$ keV for the case shown), with width parameter Δ.

$$\epsilon = \frac{2^{5/3}\sqrt{2}}{\sqrt{3}} \frac{\rho X_A X_B}{m_H^2 A_A A_B \sqrt{\mu}} Q S_0 \frac{E_G^{1/6}}{(kT)^{2/3}} \exp\left[-3\left(\frac{E_G}{4kT}\right)^{1/3}\right]. \tag{3.134}$$

Equation 3.134 can tell us, for example, the luminosity produced by the p-p chain in the Sun. For a rough estimate, let us take for the mass density in the core of the Sun the central density, $\rho = 150$ g cm^{-3}. As already noted in Section 3.6, in the central regions of the Sun, some of the hydrogen has already been converted into helium by nuclear reactions. Let us assume a typical hydrogen abundance of $X = 0.5$, which we can use for X_A and X_B. The first step in the p-p chain, the $p + p \rightarrow d + e^+ + \nu_e$ reaction, is by far the slowest of the three steps in the chain, and it is therefore the bottleneck that sets the rate of the entire p-p process. The constant S_0 for this reaction is calculated theoretically to be $\approx 4 \times 10^{-46}$ cm^2 keV, which is characteristic of weak interactions. For Q, let us take the entire thermal energy release of each p-p chain completion, since once the first step occurs, on a timescale of 10^{10} yr, the following two reactions, with timescales of order 1 s and 300,000 yr, respectively, are essentially instantaneous. We saw that every completion of the chain produces 26.73 MeV of energy and two neutrinos. Subtracting the 0.52 MeV carried off, on average, by the two neutrinos, the thermal energy released per p-p chain completion is $Q = 26.2$ MeV. As already noted, $E_G = 500$ keV for two protons, and the typical core temperature is $kT = 1$ keV. The atomic mass numbers are, of course, $A_A = A_B = 1$, and the reduced mass is $\mu = m_p/2$. Finally, since we are considering a reaction between identical particles (i.e., protons on protons) we need to divide the collision rate by 2, to avoid double counting. With these numbers, Eq. 3.134 gives a power density of

$$\epsilon = 10 \text{ erg s}^{-1} \text{g}^{-1}. \tag{3.135}$$

Multiplying this by the mass of the core of the Sun, say, $0.2M_\odot = 4 \times 10^{32}$ g, gives a luminosity of $\sim 4 \times 10^{33}$ erg s^{-1}, in good agreement with the observed solar luminosity of 3.8×10^{33} erg s^{-1}.

Reviewing the derivation of Eqs. 3.122–3.134, we see that we can also recover the reaction rate per nucleus, dN_A/dt, by dividing back from ϵ a factor $(X_A Q)/(m_H A_A)$. For the $p + p$ reaction, this gives a rate of 1.6×10^{-18} s^{-1} per proton. The reciprocal of this rate is the typical time a proton has to wait until it reacts with another proton, and indeed equals

$$\tau_{pp} \sim 6 \times 10^{17} \text{ s} \sim 2 \times 10^{10} \text{ yr}, \tag{3.136}$$

as asserted above. Thus, we have shown that hydrogen fusion provides an energy source that can power the observed luminosity of the Sun over the known age of the Solar System, about 5 billion years, not only in terms of energy budget (Eq. 3.102) but also in terms of the energy generation *rate*. Furthermore, we see that the timescale to deplete the hydrogen fuel in the solar core is of order 10 billion years.

The total power density at a point in a star with a given temperature, density, and abundance will be the sum of the power densities due to all the possible nuclear reactions, each described by 3.134. Because of the exponential term in 3.134, there will be a strong preference for reactions between species with low atomic number, and hence small E_G. For example, compare the reactions

$$p + d \rightarrow {}^3\text{He} + \gamma \qquad (E_G = 0.66 \text{ MeV}) \tag{3.137}$$

and

$$p + {}^{12}\text{C} \rightarrow {}^{13}\text{N} + \gamma \qquad (E_G = 35.5 \text{ MeV}), \tag{3.138}$$

which have comparable nuclear cross sections S_0 (both reactions follow the same process of adding a proton to a nucleus and emitting a photon). At a typical kinetic energy of 1 keV, if the abundances of deuterium and carbon nuclei were comparable, the ratio between the rates would be

$$\frac{\mathcal{R}(p^{12}\text{C})}{\mathcal{R}(pd)} \sim \exp\left[-3\frac{35.5^{1/3} - 0.66^{1/3}}{(4 \times 0.001)^{1/3}}\right] \sim e^{-46} \sim 10^{-20}. \tag{3.139}$$

Furthermore, the higher the Gamow energy, the more strongly will the reaction rate depend on temperature. For example, a first-order Taylor expansion of Eq. 3.134 around $T = 1.5 \times 10^7$ K, the central temperature of the Sun, shows that the $p + p \rightarrow d + e^+ + \nu_e$ rate depends on temperature approximately as T^4, while the $p + {}^{12}\text{C} \rightarrow {}^{13}\text{N} + \gamma$ rate goes like T^{18} (see Problem 8). The steep positive temperature dependence of nuclear reactions, combined with the virial theorem, means that nuclear reactions serve as a natural "thermostat" that keeps stars stable. Suppose, for example, that the temperature inside a star rises. This will increase the rate of nuclear reactions, leading to an increase in luminosity. Due to opacity, this additional energy will not directly escape from the star, resulting in a temporary increase in total energy. Since

$$E_{tot} = \tfrac{1}{2} E_{gr} = -E_{th}, \tag{3.140}$$

the gravitational energy E_{gr} will grow (i.e., become less negative), meaning the star will expand, and E_{th} will become smaller, meaning the temperature will be reduced again. This explains why main-sequence stars of very different masses have comparable core temperatures.

The thermostatic behavior controls also the long-term evolution of stars. Eventually, when the dominant nuclear fuel runs out, the power density ϵ will drop. The star will then contract, E_{th} will increase, and T will rise until a new nuclear reaction, involving nuclei of higher atomic number, can become effective.

A key prediction of the picture we have outlined, in which the energy of the Sun derives from the p-p chain, is that there will be a constant flux of neutrinos coming out of the Sun. As opposed to the photons, the weak interaction of the neutrinos with matter guarantees that they can escape the core of the Sun almost unobstructed.[7] As calculated above, the thermal energy released per p-p chain completion is 26.2 MeV. The neutrino number flux on Earth should therefore be twice the solar energy flux divided by 26.2 MeV:

$$f_{neutrino} = \frac{2f_\odot}{26.2 \text{ MeV}} = \frac{2 \times 1.4 \times 10^6 \text{ erg s}^{-1} \text{ cm}^{-2}}{26.2 \times 1.6 \times 10^{-6} \text{ erg}} = 6.7 \times 10^{10} \text{ s}^{-1} \text{ cm}^{-2}. \tag{3.141}$$

This huge particle flux goes mostly unhindered through our bodies and through the entire Earth, and is extremely difficult to detect. Experiments to measure the solar neutrino flux began in the 1960s, and have consistently indicated a deficit in the flux of electron neutrinos arriving from the Sun. It now appears most likely that the **total** neutrino flux from the Sun is actually very close to the predictions of solar models. The observed deficit is the result of previously unknown *flavor oscillations*, in which some of the original electron neutrinos turn into other types of neutrinos enroute from the Sun to the Earth.

We note, for completeness, that apart from the particular p-p chain described in Eqs. 3.97–3.99, which is the main nuclear reaction sequence in the Sun, other nuclear reactions occur, and produce neutrinos that are detectable on Earth (see Problem 9). In stars more massive than the Sun, hydrogen is converted to helium also via a different sequence of reactions, called the CNO cycle. In the CNO cycle, the trace amounts of carbon, nitrogen, and oxygen in the gas serve as *catalysts* in the hydrogen-to-helium burning, without any additional C, N, or O being synthesized. The main branch of the CNO cycle actually begins with reaction 3.138,

$$p + {}^{12}\text{C} \rightarrow {}^{13}\text{N} + \gamma. \tag{3.142}$$

---

[7] Typical cross sections for scattering of neutrinos on matter are of order 10^{-43} cm^2, 10^{18} times smaller than the Thomson cross section for photons. Scaling from Eq. 3.41, the mean free path for neutrinos in the Sun is $\sim 10^{18}$ cm, 10^7 times greater than the solar radius.

This is followed by

$$^{13}\text{N} \rightarrow {}^{13}\text{C} + e^+ + \nu_e, \tag{3.143}$$

$$p + {}^{13}\text{C} \rightarrow {}^{14}\text{N} + \gamma, \tag{3.144}$$

$$p + {}^{14}\text{N} \rightarrow {}^{15}\text{O} + \gamma. \tag{3.145}$$

$$^{15}\text{O} \rightarrow {}^{15}\text{N} + e^+ + \nu_e, \tag{3.146}$$

and finally

$$p + {}^{15}\text{N} \rightarrow {}^{12}\text{C} + {}^4\text{He}. \tag{3.147}$$

Although we noted that reaction 3.142 is slower by 20 orders of magnitude than the p-p chain's $p + d \rightarrow {}^3\text{He} + \gamma$, the $p + d$ reaction can take place only after overcoming the $p + p$ bottleneck, which has a timescale 18 orders of magnitude *longer* than $p + d$. The lack of such a bottleneck for the $p + {}^{12}\text{C}$ reaction is further compensated by this reaction's strong dependence on temperature. Although core temperature varies only weakly with stellar mass, the slightly higher core temperatures in more massive stars are enough to make the CNO cycle the dominant hydrogen-burning mechanism in main-sequence stars of mass $1.2 M_\odot$ and higher.

3.11 Solution of the Equations of Stellar Structure

We have now derived the four differential equations and the three additional functions that, together with boundary conditions, define uniquely the equilibrium properties of a star of a given mass and composition. Along the way, we already deduced many of the observed properties of main-sequence stars. "Solving" this system of coupled equations means finding the functions $P(r)$, $T(r)$, and $\rho(r)$, which are the ones that are usually considered to describe the structure of the star. Unfortunately, there is no analytic solution to the equations, unless some unrealistic assumptions are made (see, e.g., Problems 4 and 5). Nevertheless, a numerical solution can be obtained straightforwardly, and is the most reasonable way to proceed anyway, given the complicated nature of the functions P, κ, and ϵ when all relevant processes are included. In a numerical solution, the differentials in the equations are replaced by differences. Then, an example of one possible calculation scheme is one in which the radial structure of a star is followed shell by shell, going either outward from the center or inward from the surface.

3.12 Convection

Under certain conditions, the main means of energy transport in some regions of a star is **convection**, rather than radiative transport. Convection occurs when a volume element of material that is displaced from its equilibrium position, rather than returning to the original position, continues moving in the displacement direction. For example, if the

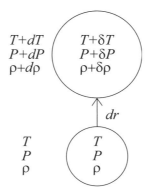

Figure 3.10 A mass element (lower circle) inside a star undergoes a small displacement dr to a higher position (upper circle), and expands adiabatically to match the new surrounding pressure $P + dP$. If, after the expansion, the density inside the element, $\rho + \delta\rho$, is larger than the surrounding density $\rho + d\rho$, the element will sink back to its former equilibrium position. If, on the other hand, the density inside the volume is lower than that of the surroundings, the mass element will be buoyed up, and convection ensues.

displacement is upward to a region of lower density, and after the displacement the density of the volume element is lower than that of its new surroundings, the element will continue to be buoyed upward. When convection sets in, it is very efficient at transporting heat, and becomes the dominant transport mechanism.

To see what are the conditions for the onset of convection, consider a volume element of gas at equilibrium radius r inside a star , where the temperature, pressure, and density are T, P, and ρ, respectively (see Fig. 3.10). Now let us displace the element to a radius $r + dr$, where the parameters of the surroundings are $T + dT$, $P + dP$, and $\rho + d\rho$. Since the gas in the star obeys

$$\rho \propto \frac{P}{T},\tag{3.148}$$

taking the logarithmic derivative gives

$$\frac{d\rho}{\rho} = \frac{dP}{P} - \frac{dT}{T}.\tag{3.149}$$

To simplify the problem, we will assume that, at its new location, the volume element expands **adiabatically** (i.e., without exchanging heat with its new surroundings, so $dQ = 0$, and therefore the entropy, defined as $dS = dQ/T$, remains constant). The element expands until its pressure matches the surrounding pressure, and reaches new parameters $T + \delta T$, $\rho + \delta\rho$, and $P + \delta P = P + dP$, where we have identified the small changes inside the element with a "δ" rather than a "d." Since we approximate the expansion of the element to be adiabatic, it obeys an equation of state

$$P \propto \rho^{\gamma},\tag{3.150}$$

where the *adiabatic index*, γ, is the usual ratio of heat capacities at constant pressure and constant volume. Taking again the logarithmic derivative, we obtain

$$\frac{\delta\rho}{\rho} = \frac{1}{\gamma}\frac{\delta P}{P}. \tag{3.151}$$

The element will continue to float up, rather than falling back to equilibrium, if after the expansion its density is lower than that of the the surroundings, i.e.,

$$\rho + \delta\rho < \rho + d\rho, \tag{3.152}$$

or simply

$$\delta\rho < d\rho, \tag{3.153}$$

(recall that both $\delta\rho$ and $d\rho$ are negative), or dividing both sides by ρ,

$$\frac{\delta\rho}{\rho} < \frac{d\rho}{\rho}. \tag{3.154}$$

Substituting from Eqs. 3.149 and 3.151, the condition for convection becomes

$$\frac{1}{\gamma}\frac{\delta P}{P} < \frac{dP}{P} - \frac{dT}{T}. \tag{3.155}$$

Recalling that $\delta P = dP$, this becomes

$$\frac{dT}{T} < \frac{\gamma-1}{\gamma}\frac{dP}{P}, \tag{3.156}$$

or upon division by dr,

$$\frac{dT}{dr} < \frac{\gamma-1}{\gamma}\frac{T}{P}\frac{dP}{dr}. \tag{3.157}$$

Since the radial temperature and pressure gradients are both negative, the condition for convection is that the temperature profile must fall fast enough with increasing radius, i.e., convection sets in when

$$\left|\frac{dT}{dr}\right| > \frac{\gamma-1}{\gamma}\frac{T}{P}\left|\frac{dP}{dr}\right|. \tag{3.158}$$

For a nonrelativistic gas without internal degrees of freedom (e.g., ionized hydrogen), $\gamma = 5/3$. As the number of internal degrees of freedom increases, γ becomes smaller, making convection possible even if $|dT/dr|$ is small. This can occur when the gas is made of atoms, which can be excited or ionized, or of molecules that have rotational and vibrational degrees of freedom and can be dissociated. Convection therefore occurs in some cool regions of stars, where atoms and molecules exist. This applies to the outer layers of intermediate-mass main-sequence stars and red giants, and to large ranges in radius in low-mass stars. Another range of applicability of convection is in the cores of massive stars. The Sun is convective in the outer 28% of its radius.

Once convection sets in, it mixes material at different radii and thus works toward equilibrating temperatures, i.e., lowering the absolute value of the temperature gradient

$|dT/dr|$. Therefore, convection can be implemented into stellar structure computations by testing, at every radius, if the convection condition has been met. If it has, convection will bring the temperature gradient back to its critical value, and therefore the radiative energy transport equation (3.51) can be replaced by Eq. 3.158, but with an equality sign:

$$\frac{dT}{dr} = \frac{\gamma - 1}{\gamma} \frac{T(r)}{P(r)} \frac{dP}{dr}. \tag{3.159}$$

Problems

1. In Eqs. 3.23–3.28, we saw that, for a star composed of a classical, nonrelativistic, ideal gas, $E_{\text{total}} = E_{\text{th}}^{\text{tot}} + E_{\text{gr}} = -E_{\text{th}}^{\text{tot}}$, and therefore the star is bound. Repeat the derivation, but for a classical **relativistic** gas of particles. Recall (Eq. 3.75) that the equation of state of a relativistic gas is $P = \frac{1}{3} E_{\text{th}}/V$. Show that, in this case, $E_{\text{gr}} = -E_{\text{th}}^{\text{tot}}$, and therefore $E_{\text{total}} = E_{\text{th}}^{\text{tot}} + E_{\text{gr}} = 0$, i.e., the star is marginally bound. As a result, stars dominated by radiation pressure are unstable.

2. The pressure inside a normal star is given by (Eq. 3.76)

   $$P = P_g + P_{\text{rad}} = \frac{\rho k T}{\bar{m}} + \frac{1}{3} a T^4.$$

   Using parameters appropriate to the Sun, show that throughout the Sun, including the core, where the internal temperature is about 10^7 K, the kinetic pressure dominates.

3. Because of the destabilizing influence of radiation pressure (see Problem 1), the most massive stars that can form are those in which the radiation pressure and the nonrelativistic kinetic pressure are approximately equal. Estimate the mass of the most massive stars, as follows.

   a. Assume that the gravitational binding energy of a star of mass M and radius R is $|E_{\text{gr}}| \sim GM^2/R$. Use the virial theorem (Eq. 3.22),

   $$\bar{P} = -\frac{1}{3} \frac{E_{\text{gr}}}{V},$$

   to show that

   $$P \sim \left(\frac{4\pi}{3^4}\right)^{1/3} GM^{2/3} \rho^{4/3},$$

   where ρ is the typical density.

   b. Show that if the radiation pressure, $P_{\text{rad}} = \frac{1}{3} a T^4$, equals the kinetic pressure, then the total pressure is

   $$P = 2 \left(\frac{3}{a}\right)^{1/3} \left(\frac{k\rho}{\bar{m}}\right)^{4/3}.$$

   c. Equate the expressions for the pressure in (a) and (b), to obtain an expression for the maximal mass of a star. Find its value, in solar masses, assuming a fully ionized hydrogen composition.

   Answer: $M = 110M_\odot$.

4. Consider a hypothetical star of radius R, with density ρ that is constant, i.e., independent of radius. The star is composed of a classical, nonrelativistic, ideal gas of fully ionized hydrogen.

   a. Solve the equations of stellar structure for the pressure profile, $P(r)$, with the boundary condition $P(R) = 0$.

   Answer: $P(r) = (2\pi/3)G\rho^2(R^2 - r^2)$.

   b. Find the temperature profile, $T(r)$.

   c. Assume that the nuclear energy production rate depends on temperature as $\epsilon \sim T^4$. (This is the approximate dependence of the rate for the p-p chain at the temperature in the core of the Sun.) At what radius does ϵ decrease to 0.1 of its central value, and what fraction of the star's volume is included within this radius?

5. Suppose a star of total mass M and radius R has a density profile $\rho = \rho_c(1 - r/R)$, where ρ_c is the central density.

   a. Find $M(r)$.

   b. Express the total mass M in terms of R and ρ_c.

   c. Solve for the pressure profile, $P(r)$, with the boundary condition $P(R) = 0$.

   Answer:

$$P(r) = \pi G \rho_c^2 R^2 \left[\frac{5}{36} - \frac{2}{3}\left(\frac{r}{R}\right)^2 + \frac{7}{9}\left(\frac{r}{R}\right)^3 - \frac{1}{4}\left(\frac{r}{R}\right)^4 \right].$$

6. Consider a star of mass $M = 10M_\odot$, composed entirely of fully ionized ^{12}C. Its core temperature is $T_c = 6 \times 10^8$ K (compared to $T_{c,\odot} = 1.5 \times 10^7$ K for the Sun).

   a. What is the mean particle mass $\bar{m}$, in units of m_H?

   Answer: 12/7.

   b. Use the classical ideal gas law, the dimensional relation between mass, density, and radius, and the virial theorem to find the scaling of the stellar radius r_* with total mass M, mean particle mass $\bar{m}$, and core temperature T_c. Using the values of these parameters for the Sun, derive the radius of the star.

   Answer: $0.70r_\odot$.

   c. If the luminosity of the star is $L = 10^7 L_\odot$, what is the effective surface temperature?

   d. Suppose the star produces energy via the reaction

$$^{12}C + {}^{12}C \rightarrow {}^{24}Mg.$$

   The atomic weight of ^{12}C is 12, and that of ^{24}Mg is 23.985. (The atomic weight of a nucleus is defined as the ratio of its mass to 1/12 the mass of a ^{12}C nucleus). What fraction of the star's mass can be converted into thermal energy?

   Answer: 6.3×10^{-4}.

e. How much time does it take for the star to use up 10% of its carbon?
   Answer: 950 yr.

7. We saw that the nuclear reaction rate in a star depends on

$$\langle \sigma v \rangle \propto \int_0^\infty f(E)dE,$$

where

$$f(E) \equiv e^{-E/kT} e^{-\sqrt{E_G/E}},$$

and E_G is the Gamow energy (Eq. 3.131).

a. By taking the derivative of $f(E)$ and equating to zero, show that $f(E)$ has a maximum at

$$E_0 = \left(\frac{kT}{2}\right)^{2/3} E_G^{1/3}.$$

b. Perform a Taylor expansion, to second order, of $f(E)$ around E_0, to approximate $f(E)$ with a Gaussian. Show that the width parameter (i.e., the "σ") of the Gaussian is

$$\Delta = \frac{2^{1/6}}{3^{1/2}} E_G^{1/6}(kT)^{5/6}.$$

Hint: Take the logarithm of $f(E)$, before Taylor expanding, and then exponentiate again the Taylor expansion.

c. Show that

$$\int_0^\infty f(E)dE = \sqrt{2\pi} f(E_0)\Delta.$$

8. Show that the dependence on temperature of the nuclear power density (Eq. 3.134) at a temperature T near T_0 can be approximated as a power law, $\epsilon \propto T^\beta$, where

$$\beta = \left(\frac{E_G}{4kT_0}\right)^{1/3} - \frac{2}{3}.$$

Evaluate β at $T_0 = 1.5 \times 10^7$ K, for the reactions $p + p \rightarrow d + e^+ + \nu_e$ and $p + {}^{12}C \rightarrow {}^{13}N + \gamma$.

Hint: From Eq. 3.134, find $\ln \epsilon$, and calculate $d(\ln \epsilon)/d(\ln T)|_{T_0}$. This is the first-order coefficient in a Taylor expansion of $\ln \epsilon$ as a function of $\ln T$ (a pure power law relation between ϵ and T would obey $\ln \epsilon = \text{const.} + \beta \ln T$).

9. We saw (Eq. 3.141) that, on Earth, the number flux of solar neutrinos from the p-p chain is

$$f_{\text{neutrino}} = \frac{2f_\odot}{26.2 \text{ MeV}} = \frac{2 \times 1.4 \times 10^6 \text{ erg s}^{-1} \text{ cm}^{-2}}{26.2 \times 1.6 \times 10^{-6} \text{ erg}}$$

$$= 6.7 \times 10^{10} \text{ s}^{-1} \text{ cm}^{-2}.$$

Other nuclear reactions in the Sun supplement this neutrino flux with a small additional flux of higher-energy neutrinos. A neutrino detector in Japan, named SuperKamiokande, consists of a tank of 50 kton of water, surrounded by photomultiplier tubes. The tubes detect the flash of **Cerenkov radiation** emitted by a recoiling electron when a high-energy neutrino scatters on it.

a. How many electrons are there in the water of the detector?

b. Calculate the detection rate for neutrino scattering, in events per day, if 10^{-6} of the solar neutrinos have a high enough energy to be detected by this experiment, and each electron poses a scattering cross section $\sigma = 10^{-43}$ cm^2.

   Hint: Consider the density of neutrino targets "seen" by an individual electron, with a relative velocity of c between the neutrinos and the electron, to obtain the rate at which one electron interacts with the incoming neutrinos, and multiply by the total number of electrons, from (a), to obtain the rate in the entire detector.

   Answers: 1.6×10^{34} electrons; 9 events per day.

4 | Stellar Evolution and Stellar Remnants

So far, we have considered only stars in static equilibrium, and found that a star of a given mass and composition has a unique, fully determined, structure. However, it is now also clear that true equilibrium cannot exist. Nuclear reactions in the central regions synthesize hydrogen into helium, and over time change the initial elemental composition. Furthermore, convection may set in at some radii and mix processed and unprocessed gas. The equations of pressure, opacity, and nuclear power density all depend sensitively on the abundances. Indeed, at some point, the hydrogen fuel in the core will be largely used up, and the star will lose the energy source that produces pressure, the gradient of which supports the star against gravitational collapse. It is therefore unavoidable that stars **evolve** with time. In this chapter, we discuss the various processes that stars of different masses undergo after the main sequence, and the properties of their compact remnants—white dwarfs, neutron stars, and black holes. We then study the phenomena that can occur when such compact objects accrete material from a companion star in a binary pair.

4.1 Stellar Evolution

Stellar evolution, as opposed to equilibrium, can be taken into account by solving a series of equilibrium stellar models (called a *stellar evolution track*), in which one updates, as a function of a star's age since formation, the gradual enrichment by elements heavier than hydrogen at different radii in the star. It turns out that the observed properties of stars on the main sequence change little during the hydrogen-burning stage, and therefore they make only small movements on the H-R diagram.

From scaling arguments, we can find the dependence of the main-sequence lifetime, $t_{\rm ms}$, on stellar mass. We previously derived the observed dependence of luminosity on mass,

$$L \sim M^{\alpha}. \tag{4.1}$$

We now also know that the energy source is nuclear reactions, whereby a fraction of a star's rest mass is converted to energy and radiated away. The total radiated energy is therefore proportional to mass,

$$Lt_{ms} \sim E \sim M, \tag{4.2}$$

and

$$t_{ms} \sim \frac{M}{L} \sim M^{1-\alpha}. \tag{4.3}$$

For intermediate-mass stars, which obey a mass–luminosity relation with $\alpha \sim 3$, $t_{ms} \sim M^{-2}$. Thus, the more massive a star, the shorter its hydrogen-burning phase on the main sequence. Detailed stellar models confirm this result. For example, the main-sequence lifetimes of stars with initial solar abundance and various masses are

$$0.5 M_\odot \rightarrow \sim 5 \times 10^{10} \text{yr};$$
$$1.0 M_\odot \rightarrow \sim 10^{10} \text{yr};$$
$$10 M_\odot \rightarrow \sim 2 \times 10^7 \text{yr}. \tag{4.4}$$

The Sun is therefore about halfway through its main-sequence lifetime. We saw that, for the most massive stars, $\alpha \sim 1$, the result of electron-scattering opacity and radiation pressure. The lifetime t_{ms} therefore becomes independent of mass and reaches a limiting value,

$$>30 M_\odot \rightarrow \sim 3 \times 10^6 \text{yr}. \tag{4.5}$$

The lifetimes of massive stars, $\sim 10^6$–10^7 yr, are short compared to the age of the Sun or the age of the Universe (which, as we will see in chapters 7–9, is about 14 gigayears [Gyr], where 1 Gyr is 10^9 yr). The fact that we observe such stars means that star formation is an ongoing process, as we will see in chapter 5.

Once most of the hydrogen in the core of a star has been converted into helium, the core contracts and the inner temperatures rise. As a result, hydrogen in the less-processed regions outside the core starts to burn in a shell surrounding the core. Stellar models consistently predict that at this stage there is a huge expansion of the outer layers of the star. The increase in luminosity, due to the gravitational contraction and the hydrogen shell burning, moves the star up in the H-R diagram, while the increase in radius lowers the effective temperature, moving the star to the right on the diagram (see Fig. 4.1). This is the **red giant** phase. The huge expansion of the star's envelope is difficult to explain by means of some simple and intuitive argument, but it is well understood and predicted robustly by the equations of stellar structure. The red-giant phase is brief compared to the main sequence, lasting roughly one-tenth the time, from a billion years for solar-mass stars, to only of order a million years for $\sim 10\, M_\odot$ stars, and a few 10^5 years for the most massive stars.

As the red-giant phase progresses, the helium core contracts and heats up, while additional helium "ash" is deposited on it by the hydrogen-burning shell. At some point,

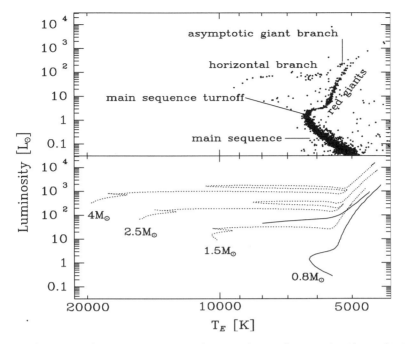

Figure 4.1 Illustration of post-main-sequence evolution on the H-R diagram. *Top:* Observed H-R diagram for stars in the *globular cluster* M3 (more on star clusters in section 5.1.5). The **main-sequence turnoff** marks the point at which stars are now leaving the main sequence and evolving on to the red-giant branch. All the stars in the cluster formed together about 13 Gyr ago, and the cluster has not experienced subsequent star formation. As a result, all stars above a certain mass, corresponding to the turnoff point, have left the main sequence, while those below that mass are still on the main sequence. The density of points in each region of the diagram reflects the amount of time spent by stars in each post-main-sequence evolution stage. *Bottom:* Theoretical stellar evolution tracks for stars of various initial main-sequence masses (with an assumed initial metal abundance of $Z = 0.0004$). Each track begins at the lower left end on the *zero-age main sequence*. After leaving the main sequence, stars evolve along, and up to the tip of, the red-giant branch. They then move quickly on the diagram to the left edge of the *horizontal branch*, where helium core burning and hydrogen shell burning take place, and evolve to the right along the horizontal branch. Once all the helium in the core has been converted to carbon and oxygen, the star rises up the "asymptotic giant branch" where double shell burning—a helium-burning shell within a hydrogen burning shell—takes place. Note the good correspondence between the theoretical track for the $0.8M_\odot$ initial-mass star (solid line) and the observed H-R diagram on top. For clarity, the theoretical horizontal and asymptotic giant branches are not shown for the other initial masses. Data credits: S.-C. Rey et al. 2001, *Astrophys. J.*, 122, 3219; and L. Girardi, et al. 2000, *Astron. Astrophys. Suppl.*, 141, 371.

the core will reach a temperature of about $T \sim 10^8$ K and a density $\rho \sim 10^4$ g cm^{-3}, where helium burning can become effective through the **triple-alpha** reaction,

$$^4\text{He} + {}^4\text{He} + {}^4\text{He} \rightarrow {}^{12}\text{C} + \gamma\,(7.275\,\text{MeV}). \tag{4.6}$$

Triple-alpha is the only reaction that can produce elements heavier than helium in the presence of only hydrogen and helium, because no stable elements exist with atomic mass numbers of 5 or 8. The beryllium isotope ^{8}Be, formed from the fusion of two ^{4}He nuclei, has a lifetime of only $\sim 10^{-16}$ s. Nevertheless, a small equilibrium abundance of ^{8}Be can be established, and capture of another ^{4}He nucleus then completes the triple-alpha process. The last stage would have an extremely low probability, were it not for the existence of an excited nuclear energy level in ^{12}C, which, when added to the rest mass energy of ^{12}C, happens to have almost exactly the rest mass energies of ^{4}He $+^8$Be. This *resonance* greatly increases the cross section for the second stage of the reaction. In fact, from the existence of abundant carbon in the Universe (without which, of course, carbon-based life would be impossible) Hoyle predicted the existence of this excited level of ^{12}C before it was discovered experimentally.

Along with carbon production, some oxygen and neon can also be synthesized via the reactions

$$^4\text{He} +^{12}\text{C} \rightarrow ^{16}\text{O} + \gamma \tag{4.7}$$

and

$$^4\text{He} +^{16}\text{O} \rightarrow ^{20}\text{Ne} + \gamma. \tag{4.8}$$

At the same time, hydrogen continues to burn in a shell surrounding the core. When helium ignition begins, the star moves quickly on the H-R diagram to the left side of the **horizontal branch**, and then evolves more slowly to the right along this branch, as seen in Fig. 4.1. Horizontal branch evolution last only about 1% of the main-sequence lifetime. Once the helium in the core has been exhausted, the core (now composed mainly of oxygen and carbon) contracts again, until a surrounding shell of helium ignites, with a hydrogen-burning shell around it. During this brief ($\sim 10^7$ yr) double-shell-burning stage, the star ascends the **asymptotic giant branch** of the H-R diagram—essentially a repeat of the red-giant branch evolution, but with helium + hydrogen shell burning around an inert carbon/oxygen core, rather than hydrogen shell burning around an inert helium core.

Evolved stars undergo large mass loss, especially on the red-giant branch and on the asymptotic giant branch, as a result of the low gravity in their extended outer regions and the radiation pressure produced by their large luminosities. Mass loss is particularly severe on the asymptotic giant branch during so-called **thermal pulses**—roughly periodic flashes of enhanced helium shell burning. These mass outflows, or stellar **winds**, lead to mass-loss rates of up to $10^{-4} M_\odot$ yr^{-1}, which rid a star of a large fraction of its initial mass. Giants are highly convective throughout their volumes, leading to a **dredge-up** of newly synthesized elements from the core to the outer layers, where they are expelled with the winds. In these processes, and additional ones we will see below, the nuclear reactions inside post-main-sequence stars create essentially all elements in the Universe that are heavier than helium.

In stars with an initial mass of less than about $8M_\odot$, as the giant phase progresses, the dense matter in the core reaches equilibrium in a new state of matter called a degenerate electron gas. As we will see in the next section, regions of the core that are in this state are supported against further gravitational contraction, even in the absence of nuclear reactions. As a result, the cores of such stars do not heat up to the temperatures required for the synthesis of heavier elements, and at the end of the asymptotic giant phase they remain with a helium/carbon/oxygen core.

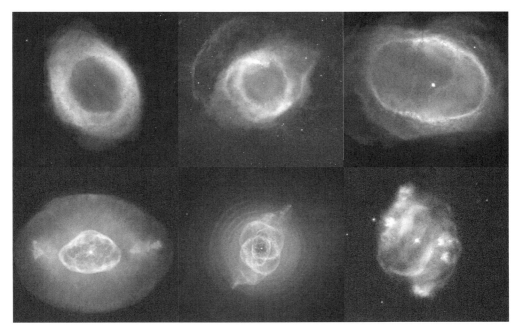

Figure 4.2 Several examples of planetary nebulae, newly formed white dwarfs that irradiate the shells of gas that were previously shed in the final stages of stellar evolution. The shells have diameters of ≈ 0.2–1 pc. Photo credits: M. Meixner, T. A. Rector, B. Balick et al., H. Bond, R. Ciardullo, NASA, NOAO, ESA, and the Hubble Heritage Team.

At this point, the remaining outer envelopes of the star expand to the point that they are completely blown off and dispersed. During this very brief stage ($\sim 10^4$ yr), the star is a **planetary nebula**[1] (see Fig. 4.2), in which ultraviolet photons from the hot, newly exposed, core excite the expanding shells of gas that previously constituted the outer layers of the star. Finally, the exposed remnant of the original core, called a "white dwarf," reaches the endpoint of stellar evolution for stars of this mass. In the white-dwarf region of the H-R diagram, these stars move with time to lower temperature and luminosity as they slowly radiate away their heat. White dwarfs are the subject of the next section.

Stars with an initial mass greater than about $8M_\odot$ continue the sequence of core contraction and synthesis of progressively heavier elements, which eventually (and quickly) ends in a supernova explosion. We shall return to this class of stars in section 4.3.

4.2 White Dwarfs

In the 19th and early 20th centuries, it was discovered that the nearby (2.7 pc) A-type star Sirius, the brightest star in the sky, is a visual binary, with a white dwarf companion that was named Sirius B. (In fact, Sirius B is the nearest known white dwarf, and was the first

---

[1] Planetary nebulae have nothing to do with planets, and the name has a purely historical origin.

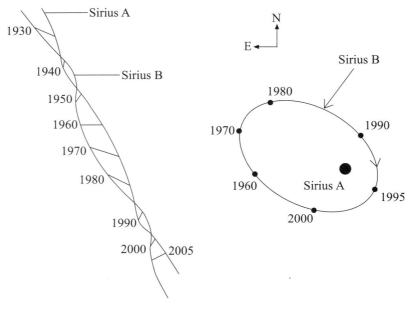

Figure 4.3 Observed motion on the sky, over the past century, of the visual binary consisting of Sirius A and its faint white dwarf companion, Sirius B. On the left are the observed positions due the orbital motions around the center of mass, combined with the proper motion of the system as a whole. On the right side, only the positions of Sirius B relative to Sirius A are shown. The maximum projected separation of the pair is 10 arcseconds. Using Kepler's law, a mass close to $1M_\odot$ is derived for the white dwarf.

one ever found.) An orbital period of about 50 years was observed (see Fig. 4.3), allowing the first measurement of the mass of a white dwarf, which turned out to be close to $1M_\odot$. Like all white dwarfs, Sirius B's low luminosity and high temperature imply a small radius of about 6000 km, i.e., less than that of the Earth. The mean density inside Sirius B is therefore of order 1 ton cm^{-3}. In this section, we work out the basic physics of white dwarfs and of matter at these extremely high densities.

4.2.1 Matter at Quantum Densities

We saw in the previous section that when the core of a star exhausts its nuclear energy supply, it contracts and heats up until reaching the ignition temperature of the next available nuclear reaction, and so on. After each contraction, the density of the core increases. At some point, the distances between atoms will be smaller than their de Broglie wavelengths. At that point, our previous assumption of a classical (rather than quantum) ideal gas, which we used to derive the equation of state, becomes invalid. To get an idea of the conditions under which this happens, recall that the de Broglie wavelength of a particle of momentum p is

$$\lambda = \frac{h}{p} = \frac{h}{(2mE)^{1/2}} \approx \frac{h}{(3mkT)^{1/2}}, \tag{4.9}$$

where we have represented the energy with the mean energy of a particle, $E \sim 3kT/2$. Since electrons and protons share the same energy, but the mass of the electron is much smaller than the mass of the proton or of other nuclei, the wavelengths of the electrons are longer, and it is the electron density that will first reach the quantum domain. At interparticle separations of order less than half a de Broglie wavelength, quantum effects should become important, corresponding to a density of

$$\rho_q \approx \frac{m_p}{(\lambda/2)^3} = \frac{8m_p(3m_e kT)^{3/2}}{h^3}. \tag{4.10}$$

For example, for the conditions at the center of the Sun, $T = 15 \times 10^6$ K, we obtain

$$\rho_q \approx \frac{8 \times 1.7 \times 10^{-24} \text{ g } (3 \times 9 \times 10^{-28} \text{ g} \times 1.4 \times 10^{-16} \text{ erg K}^{-1} \times 15 \times 10^6 \text{K})^{3/2}}{(6.6 \times 10^{-27} \text{ erg s})^3}$$

$$= 640 \text{ g cm}^{-3} \tag{4.11}$$

The central density in the Sun is $\rho \approx 150$ g cm^{-3}, and thus the gas in the Sun is still in the classical regime. Even very dense gas can remain classical, if it is hot enough. For example, for $T = 10^8$ K, i.e., $E \sim kT \sim 10$ keV,

$$\rho_q \approx 11,000 \text{ g cm}^{-3}. \tag{4.12}$$

Instead of the Maxwell-Boltzmann distribution, the energy distribution at quantum densities will follow Bose-Einstein statistics for bosons (particles with spin that is an integer multiple of $\hbar$) or Fermi-Dirac statistics for fermions (particles with spin that is an uneven integer multiple of $\hbar/2$). Let us develop the equation of state for such conditions.

4.2.2 Equation of State of a Degenerate Electron Gas

Heisenberg's uncertainty principle states that, due to the wave nature of matter, the position and momentum of a particle are simultaneously defined only to within an uncertainty

$$\Delta x \Delta p_x > h. \tag{4.13}$$

Similar relations can be written for each of the coordinates, x, y, and z. Multiplying the relations, we obtain

$$\Delta x \Delta y \Delta z \Delta p_x \Delta p_y \Delta p_z > h^3, \tag{4.14}$$

or

$$d^3\mathbf{p} dV > h^3. \tag{4.15}$$

The constant h^3 thus defines the six-dimensional volume of a "cell" in position–momentum phase space. The uncertainty principle implies that two identical particles

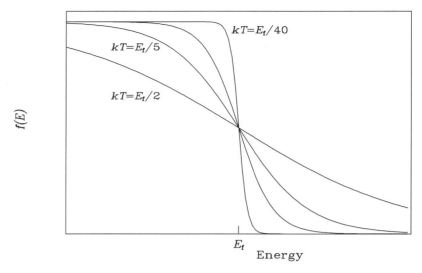

Figure 4.4 Approach to degeneracy of the Fermi-Dirac occupation number, $f(E)$, as $kT \to 0$, shown for $kT = E_f/2$, $E_f/5$, $E_f/10$, and $E_f/40$. At $kT \ll E_f$, all particles occupy the lowest energy state possible without violating the Pauli exclusion principle. The distribution then approaches a step function, with all energy states below E_f occupied, and all those above E_f empty.

that are in the same phase–space cell are in the same quantum state. According to Pauli's exclusion principle, two identical fermions cannot occupy the same quantum state. Thus, fermions that are closely packed and hence localized into a small volume, dV, must each have a large uncertainty in momentum, *and* have momenta **p** that are different from those of the other fermions in the volume. This necessarily pushes the fermions to large **p**'s, and large momenta mean large pressure.

The **Fermi-Dirac phase–space distribution**, embodying these principles for an ideal gas of fermions, is

$$dN = \frac{2s + 1}{\exp\left(\frac{E - \mu(T)}{kT}\right) + 1} \frac{d^3\mathbf{p}\,dV}{h^3}, \tag{4.16}$$

where s is the spin of each fermion in units of $\hbar$, and $\mu(T)$ is the chemical potential[2] of the gas. When $T \to 0$, then $\mu(T)$ approaches an asymptotic value, E_f. When $kT \ll E_f$, the first term in the Fermi-Dirac distribution (the *occupation number*) approaches a step function (see Fig. 4.4) in which all particles occupy the lowest energy states possible without violating the Pauli principle. This means that all energy states up to an energy E_f are occupied, and all above E_f are empty. Under such conditions, the gas is said to be **degenerate**. For

---

[2] The chemical potential of a thermodynamic system is the change in energy due to the introduction of an additional particle, at constant entropy and volume.

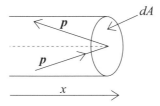

Figure 4.5 Calculation of the pressure exerted by particles of an ideal gas with momentum **p** that are reflected off the side of a container.

degenerate electrons, which are $s = 1/2$ particles, having an isotropic velocity field, the phase–space distribution will be

$$dN(p)dp = \begin{cases} 2 \times 4\pi p^2 \frac{dp\, dV}{h^3} & \text{if } |\mathbf{p}| \le p_f \\ 0 & \text{if } |\mathbf{p}| > p_f \end{cases}, \tag{4.17}$$

where p_f, called the **Fermi momentum**, is the magnitude of the momentum corresponding to the **Fermi energy** E_f. Dividing by dV, we obtain the number density of electrons of a given momentum p:

$$n_e(p)dp = \begin{cases} 8\pi p^2 \frac{dp}{h^3} & \text{if } |\mathbf{p}| \le p_f \\ 0 & \text{if } |\mathbf{p}| > p_f \end{cases}. \tag{4.18}$$

Integrating over all momenta from 0 to p_f gives a relation between the electron density and p_f:

$$n_e = \int_0^{p_f} \frac{8\pi}{h^3} p^2 dp = \frac{8\pi}{3h^3} p_f^3. \tag{4.19}$$

Next, let us derive a general expression for the pressure exerted by any ideal gas. By definition, an ideal gas consists of particles that interact only at short distances, and hence can transfer momentum only during an "impact" with another particle. Consider ideal gas particles impinging on the side of a container, with a mean interval dt between consecutive impacts (see Fig. 4.5). Set the x axis perpendicular to the surface. Particles with an x component of momentum p_x will transfer a momentum $2p_x$ to the surface with each reflection. The force per unit area due to each collision is then

$$\frac{dF_x}{dA} = \frac{2p_x}{dA\, dt} = \frac{2p_x v_x}{dA\, dx} = \frac{2p_x v_x}{dV}, \tag{4.20}$$

where $v_x = dx/dt$. The pressure is obtained by summing the forces due to all particles of all momenta:

$$P = \int_0^\infty dN(p) \frac{p_x v_x}{dV} dp, \tag{4.21}$$

where we have divided by 2 because, at any given time, only half of all the particles will have a v_x component toward the side of the container, rather than away from it. But

$$p_x v_x = m v_x^2 = \tfrac{1}{3} m v^2 = \tfrac{1}{3} p v, \tag{4.22}$$

where we have utilized $v_x^2 + v_y^2 + v_z^2 = v^2$ and assumed that the velocities are isotropic so that, on average, $v_x^2 = v_y^2 = v_z^2$. Since $dN/dV \equiv n$,

$$P = \frac{1}{3} \int_0^\infty n(p) p v \, dp. \tag{4.23}$$

Replacing the Maxwell-Boltzmann distribution for $n(p)$ recovers the classical equation of state,

$$P = nkT. \tag{4.24}$$

For a nonrelativistic[3] **degenerate** electron gas, however, we replace $n(p)$ with the Fermi-Dirac distribution in the degenerate limit (Eq. 4.18). Taking $v = p/m_e$, we obtain instead

$$P_e = \frac{1}{3} \int_0^{p_f} \frac{8\pi}{h^3} \frac{p^4}{m_e} dp = \frac{8\pi}{3 h^3 m_e} \frac{p_f^5}{5} = \left(\frac{3}{8\pi}\right)^{2/3} \frac{h^2}{5 m_e} n_e^{5/3}, \tag{4.25}$$

where we have used Eq. 4.19 to express p_f in terms of n_e. To relate n_e to the mass density appearing in the equations of stellar structure, consider a fully ionized gas composed of a particular element, of atomic number $\mathcal{Z}$ and atomic mass number A, and a density of ions n_+. Then

$$n_e = \mathcal{Z} n_+ = \mathcal{Z} \frac{\rho}{A m_p}. \tag{4.26}$$

Substituting into Eq. 4.25, we obtain a useful form for the **equation of state of a degenerate nonrelativistic electron gas**:

$$P_e = \left(\frac{3}{\pi}\right)^{2/3} \frac{h^2}{20 m_e m_p^{5/3}} \left(\frac{\mathcal{Z}}{A}\right)^{5/3} \rho^{5/3}. \tag{4.27}$$

The important feature of this equation of state is that the electron pressure does not depend on temperature. Indeed, in our derivation of this equation, we have assumed that kT is effectively zero. (More precisely, kT is very low compared to the energy of the most energetic electrons at the Fermi energy, which are prevented from occupying lower energy states by the Pauli principle—see Problem 1.)

In a typical white dwarf, $\rho \sim 10^6$ g cm^{-3} and $T \sim 10^7$ K. White dwarfs are generally composed of material that was processed by nuclear reactions into helium, carbon, and oxygen, and therefore $\mathcal{Z}/A \approx 0.5$. Plugging these numbers into 4.27, we find

---

[3] Note that, although we have used nonrelativistic considerations (Eq. 4.22) to derive Eq. 4.23, it holds in the relativistic case as well. We can easily verify that, for an ultrarelativistic gas with particle energies E, by replacing p with E/c, v with c, and $n(p)dp$ with $n(E)dE$, we recover the relation $P = u/3$, which we derived in Eq. 3.74.

$$P_e \sim \frac{(6.6 \times 10^{-27} \text{ erg s})^2}{20 \times 9 \times 10^{-28} \text{ g } (1.7 \times 10^{-24} \text{ g})^{5/3}} 0.5^{5/3} (10^6 \text{ g cm}^{-3})^{5/3}$$

$$= 3 \times 10^{22} \text{ dyne cm}^{-2}. \tag{4.28}$$

As opposed to the electrons, the nuclei at such densities are still completely in the classical regime. The thermal pressure due to the nuclei, assuming a helium composition, is

$$P_{\text{th}} = n_+ kT = \frac{\rho}{4m_p} kT \sim \frac{10^6 \text{ g cm}^{-3} \times 1.4 \times 10^{-16} \text{ erg K}^{-1} \times 10^7 \text{ K}}{4 \times 1.7 \times 10^{-24} \text{ g}}$$

$$= 2 \times 10^{20} \text{dyne cm}^{-2}. \tag{4.29}$$

The degenerate electron pressure therefore completely dominates the pressure in the star.

4.2.3 Properties of White Dwarfs

Next, we can see what the degenerate electron pressure equation of state, combined with the other equations of stellar structure, implies for the properties of white dwarfs. Let us start with the relation between mass and radius.

4.2.3.1 Mass–Radius Relationship
The equations of mass continuity and hydrostatic equilibrium, expressed as scaling relations (see Eqs. 3.80 and 3.81), suggest

$$\rho \sim \frac{M}{r^3}, \tag{4.30}$$

and

$$P \sim \frac{GM\rho}{r} \sim \frac{GM^2}{r^4}. \tag{4.31}$$

The degenerate electron-gas equation of state is

$$P \sim b\rho^{5/3} \sim b\frac{M^{5/3}}{r^5}, \tag{4.32}$$

where the constant factor b is given in Eq. 4.27. Equating the pressures gives

$$r \sim \frac{b}{G} M^{-1/3}. \tag{4.33}$$

In other words, the radius of a white dwarf decreases with increasing mass. An order-of-magnitude estimate of the radius is therefore

$$r_{\text{wd}} \sim \frac{b}{G} M^{-1/3} \sim \frac{h^2}{20 m_e m_p^{5/3} G} \left(\frac{Z}{A}\right)^{5/3} M^{-1/3}$$

$$\sim \frac{(6.6 \times 10^{-27} \text{ erg s})^2 (2 \times 10^{33} \text{ g})^{-1/3}}{20 \times 9 \times 10^{-28} \text{ g } (1.7 \times 10^{-24} \text{ g})^{5/3} \, 6.7 \times 10^{-8} \text{ cgs}} \left(\frac{Z}{A}\right)^{5/3} \left(\frac{M}{M_\odot}\right)^{-1/3}$$

$$= 1.2 \times 10^9 \text{cm} \left(\frac{Z}{A}\right)^{5/3} \left(\frac{M}{M_\odot}\right)^{-1/3}, \tag{4.34}$$

i.e., about 4000 km for $\mathcal{Z}/A = 0.5$ and $M = 1 M_\odot$, as deduced for observed white dwarfs from their luminosities and temperatures. A full solution of the equations of stellar structure for the degenerate gas equation of state gives

$$r_{\text{wd}} = 2.3 \times 10^9 \text{cm} \left(\frac{\mathcal{Z}}{A} \right)^{5/3} \left(\frac{M}{M_\odot} \right)^{-1/3}. \tag{4.35}$$

4.2.3.2 The Chandrasekhar Mass

The larger the white-dwarf mass that we consider, the smaller r_{wd} becomes, implying larger densities, and therefore larger momenta to which the electrons are pushed. When the electron velocities become comparable to the speed of light, we can no longer assume $v = p/m$ in Eq. 4.23. Instead, v, which dictates the rate at which collisions transfer momentum to the container wall, approaches c. In the *ultrarelativistic* limit, we can replace v with c. Equation 4.25 is then replaced with

$$P_e = \frac{1}{3} \int_0^{p_f} \frac{8\pi}{h^3} p^2 pc \, dp = \frac{8\pi c}{3h^3} \frac{p_f^4}{4}. \tag{4.36}$$

Again using Eqs. 4.19 and 4.26, we obtain the **equation of state for an ultrarelativistic degenerate spin-1/2 fermion gas**:

$$P_e = \left(\frac{3}{8\pi} \right)^{1/3} \frac{hc}{4m_p^{4/3}} \left(\frac{\mathcal{Z}}{A} \right)^{4/3} \rho^{4/3}. \tag{4.37}$$

Compared to the nonrelativistic case (Eq. 4.27), note the 4/3 power, but also the fact that the electron mass does not appear, i.e., this equation holds for *any* ultrarelativistic degenerate ideal gas of spin-1/2 particles. This comes about because, for ultrarelativistic particles, the rest mass is a negligible fraction of the total energy, $E = (m^2 c^4 + p^2 c^2)^{1/2}$, and hence $p \approx E/c$. As we go from small to large white-dwarf masses there will be a gradual transition from the nonrelativistic to the ultrarelativistic equation of state, with the power-law index of ρ gradually decreasing from 5/3 to 4/3.

This necessarily means that, as we go to higher masses, and the density increases due to the shrinking radius, the pressure support will rise more and more slowly, so that the radius shrinks even more sharply with increasing mass.[4] To see what happens as a result, let us rederive the scaling relations between mass and radius, but with an index $(4 + \epsilon)/3$, and then let ϵ approach 0. Thus,

$$P \sim \rho^{(4+\epsilon)/3}, \tag{4.38}$$

so

$$\frac{M^{(4+\epsilon)/3}}{r^{4+\epsilon}} \sim \frac{M^2}{r^4}, \tag{4.39}$$

---

[4] Sirius B, with a mass of $1 M_\odot$, is among the more massive white dwarfs known, and its equation of state is already in the mildly relativistic regime. Its radius, 5880 km, is therefore smaller than would be expected based on Eq 4.35, but is fully consistent with the results of a relativistic calculation.

or

$$r^{\epsilon} \sim M^{(\epsilon-2)/3}, \tag{4.40}$$

and

$$r \sim M^{(\epsilon-2)/3\epsilon}. \tag{4.41}$$

When $\epsilon \to 0$,

$$r \to M^{-\infty} = 0. \tag{4.42}$$

In other words, at a mass high enough so that the electrons become ultrarelativistic, the electron pressure becomes incapable of supporting the star against gravity, the radius shrinks to zero (and the density rises to infinity), unless some other source of pressure sets in. We will see that, at high enough density, the degeneracy pressure due to protons and neutrons begins to operate, and it can sometimes stop the full gravitational collapse, producing objects called neutron stars.

The above argument implies that there is a maximum stellar mass that can be supported by degenerate electron pressure. It is called the **Chandrasekhar mass**. To estimate it, recall from the virial theorem that

$$\bar{P}V = -\tfrac{1}{3}E_{gr}. \tag{4.43}$$

Substituting the ultrarelativistic electron degeneracy pressure for $\bar{P}$, and the usual expression for the self-energy E_{gr}, we can write

$$\left(\frac{3}{8\pi}\right)^{1/3} \frac{hc}{4m_p^{4/3}} \left(\frac{Z}{A}\right)^{4/3} \rho^{4/3} V \sim \frac{1}{3} \frac{GM^2}{r}. \tag{4.44}$$

With

$$\rho \sim \frac{M}{V} \tag{4.45}$$

and

$$V = \frac{4\pi}{3} r^3, \tag{4.46}$$

r cancels out of the equation and we obtain

$$M \sim 0.11 \left(\frac{Z}{A}\right)^2 \left(\frac{hc}{Gm_p^2}\right)^{3/2} m_p. \tag{4.47}$$

A full solution of the equations of stellar structure for this equation of state gives a somewhat larger numerical coefficient, so that the Chandrasekhar mass is

$$M_{ch} = 0.21 \left(\frac{Z}{A}\right)^2 \left(\frac{hc}{Gm_p^2}\right)^{3/2} m_p. \tag{4.48}$$

The expression $Gm_p^2/(hc)$ that appears in the Chandrasekhar mass is a dimensionless constant that can be formed by taking a proton's gravitational self-energy, with the proton radius expressed by its de Broglie wavelength, and forming the ratio with the proton's rest energy:

$$\alpha_G \equiv \frac{Gm_p^2}{h/(m_p c)m_p c^2} = \frac{Gm_p^2}{hc}$$

$$= \frac{6.7 \times 10^{-8} \text{ cgs } (1.7 \times 10^{-24} \text{ g})^2}{6.6 \times 10^{-27} \text{ erg s} \times 3 \times 10^{10} \text{ cm s}^{-1}} \approx 10^{-39}. \tag{4.49}$$

The constant α_G expresses the strength of the gravitational interaction, and is the gravitational analog of the fine-structure constant,

$$\alpha_{\text{em}} = \frac{e^2}{\hbar c} \approx \frac{1}{137}, \tag{4.50}$$

which expresses the strength of the electromagnetic interaction. Equation 4.48 says that the maximum mass of a star supported by electron degeneracy pressure is, to an order of magnitude, the mass of $\alpha_G^{-3/2}$ protons (i.e., $\sim 10^{57}$ protons). Since $\mathcal{Z}/A \approx 0.5$,

$$M_{\text{ch}} = 0.21 \times 0.5^2 \times 10^{39\frac{3}{2}} \times 1.7 \times 10^{-24} \text{ g} = 1.4 M_\odot. \tag{4.51}$$

In fact, no white dwarfs with masses higher than M_{ch} have been found.

There is also a lower bound to the masses of isolated[5] white dwarfs that have been measured, of about $0.25 M_\odot$. This, however, is a result of the finite age of the Universe, 1.4×10^{10} yr. Stars that will form white dwarfs having masses smaller than this (namely, stars that have an initial mass on the main sequence smaller than about $0.8 M_\odot$) have not yet had time to go through their main-sequence lifetimes, even if they were formed early in the history of the Universe.

4.2.3.3 White Dwarf Cooling

Due to the good thermal conduction of the degenerate electrons in a white dwarf (similar to the conduction in metals, which arises in the same way), the temperature inside a white dwarf is approximately constant with radius. The temperature can be estimated by recalling that a white dwarf forms from the contraction of a thermally unsupported stellar core, of mass M, down to the radius at which degeneracy pressure stops the contraction. Just before reaching that final point of equilibrium, from the virial theorem, the thermal energy will equal half the gravitational energy:

$$E_{\text{th}} \sim \frac{1}{2} \frac{GM^2}{r}. \tag{4.52}$$

---

[5] In interacting binaries, ablation by beams of matter and radiation from a companion can sometimes lower the mass of a white dwarf, or even destroy the white dwarf completely. See section 4.6.3.

For a pure helium composition, the number of nuclei in the core is $M/4m_H$, and the number of electrons is $M/2m_H$. The total thermal energy (which, once degeneracy sets it, will no longer play a role in supporting the star against gravity) is therefore

$$E_{th} = \frac{3}{2}NkT = \frac{3}{2}\frac{M}{m_p}\left(\frac{1}{2}+\frac{1}{4}\right)kT = \frac{9}{8}\frac{M}{m_p}kT, \tag{4.53}$$

and so

$$kT \sim \frac{4}{9}\frac{GMm_p}{r}. \tag{4.54}$$

Substituting the equilibrium r_{wd} of white dwarfs from Eq. 4.34 yields

$$kT \sim \frac{80G^2 m_e m_p^{8/3}}{9h^2}\left(\frac{Z}{A}\right)^{-5/3}M^{4/3}$$

$$= \frac{80(6.7\times10^{-8}\text{ cgs})^2\,9\times10^{-28}\text{ g }(1.7\times10^{-24}\text{ g})^{8/3}}{9(6.6\times10^{-27}\text{ erg s}^{-1})^2}\,0.5^{-5/3}(1\times10^{33}\text{ g})^{4/3}$$

$$= 1.1\times10^{-8}\text{erg}, \tag{4.55}$$

for a $0.5M_\odot$ white dwarf. The temperature is thus $kT \sim 70$ keV, or $T \sim 8\times10^8$ K, and a just-formed degenerate core is a very hot object, with thermal emission that peaks in the X-ray part of the spectrum. As such, once the core becomes an exposed white dwarf, its radiation ionizes the layers of gas that were blown off in the various stages on the giant phase. As already noted, this produces the objects called planetary nebulae.

A white dwarf is an endpoint in stellar evolution, devoid of nuclear reactions. It therefore cools by radiating from its surface the thermal energy stored in the still-classical gas of nuclei within the star's volume. (The degeneracy of the electron gas limits almost completely the ability of the electrons to lose their kinetic energies.) The radiated luminosity will be

$$L = 4\pi r_{wd}^2\sigma T_E^4, \tag{4.56}$$

where T_E is the effective temperature of the white-dwarf photosphere. Although electron heat conduction leads to a constant temperature over most of the volume, there is a thin nondegenerate surface layer (of order 1% of the white-dwarf radius) that insulates the star. This layer lowers T_E relative to the interior temperature and slows down the rate of energy loss.

However, to obtain a crude upper limit on the rate at which a white dwarf cools by means of its radiative energy loss, let us assume a constant temperature all the way out to the surface of the star, so that $T_E \sim T$. The radiative energy loss rate is then

$$4\pi r_{wd}^2\sigma T^4 \sim \frac{dE_{th}}{dt} = \frac{3Mk}{8m_p}\frac{dT}{dt} \tag{4.57}$$

(where we have included in the right-hand term only the contribution of the nuclei to the thermal energy from Eq. 4.53). Separating the variables T and t, and integrating, the cooling time to a temperature T is

$$
\begin{aligned}
\tau_{\text{cool}} &\sim \frac{3Mk}{8m_p 4\pi r_{\text{wd}}^2 \sigma 3T^3} \\
&= \frac{3 \times 1 \times 10^{33} \text{ g} \times 1.4 \times 10^{-16} \text{ erg K}^{-1}}{8 \times 1.7 \times 10^{-24} \text{ g } 4\pi (4 \times 10^8 \text{ cm})^2 \times 5.7 \times 10^{-5} \text{ cgs} \times 3T^3} \\
&= 3 \times 10^9 \text{yr} \left(\frac{T}{10^3 \text{ K}}\right)^{-3},
\end{aligned}
\tag{4.58}
$$

where we have taken $M = 0.5 M_\odot$ and $r_{\text{wd}} = 4000$ km. (We have abbreviated the units of the Stefan-Boltzmann law's σ as cgs.) Alternatively, we can write the temperature as a function of time as

$$
\frac{T}{10^3 \text{ K}} \sim \left(\frac{t}{3 \times 10^9 \text{ yr}}\right)^{-1/3}
\tag{4.59}
$$

Thus, even with the unrealistically efficient cooling we have assumed, it would take a $0.5 M_\odot$ white dwarf several gigayears to cool to 10^3 K. In reality, the insulation of the nondegenerate surface layer results in an effective temperature that is significantly lower than the interior temperature, and hence an even lower cooling rate. Furthermore, at some point in the cooling evolution, crystallization of the nucleons inside the white dwarf takes place, and the latent heat that is released and added to the thermal balance further slows down the decline in temperature. Detailed models have been calculated that take these and other processes into account for various masses and chemical compositions of white dwarfs (a carbon/oxygen core is usually assumed, surrounded by helium and hydrogen envelopes). The models show that over 10^{10} yr, comparable to the age of the Universe, white dwarfs cannot cool below $\sim$3000–4000K. This explains why most white dwarfs are observed to have high temperatures, and hence their blue to white colors. The coolest white dwarfs known have effective temperatures of $\sim$3500 K.

4.2.3.4 Brown Dwarfs

Let us digress for a moment from the subject of stellar remnants, and use the equations we have developed to see that electron degeneracy and its consequences also dictate a minimal initial mass that a star must have to shine. Consider a newly forming star (or "protostar") composed of a collapsing cloud of hydrogen. Nuclear ignition of hydrogen requires a minimal temperature of about $T_{\text{ign}} \approx 10^7$ K. Recall the relation between temperature and mass of a white dwarf, obtained by arguing that the contraction of the core will halt at the radius when degeneracy pressure sets in (Eq. 4.55). However, for fully ionized hydrogen (as opposed to helium), $N = 2M/m_p$, (rather than $N = 3M/4m_p$; Eq. 4.53). There are 8/3 times more particles, and the temperature is correspondingly lower, so Eq. 4.55 becomes

$$
kT \sim \frac{10G^2 m_e m_p^{8/3}}{3h^2} \left(\frac{\mathcal{Z}}{A}\right)^{-5/3} M^{4/3}.
\tag{4.60}
$$

If the mass of the protostar is small enough such that contraction halts before T_{ign} is attained, the object will never achieve true stardom on the main sequence. The limiting mass is

$$M_{\text{min}} \sim (kT_{\text{ign}})^{3/4} \left(\frac{10G^2 m_e m_p^{8/3}}{3h^2} \right)^{-3/4} \left(\frac{\mathcal{Z}}{A} \right)^{5/3} = 0.09 M_\odot, \tag{4.61}$$

where we have assumed $\mathcal{Z}/A = 1$, appropriate for hydrogen. A full solution of the stellar structure equations gives

$$M_{\text{min}} \approx 0.07 M_\odot. \tag{4.62}$$

Such "failed stars," with masses lower than this limit, are called **brown dwarfs**. As noted in section 2.2.2, stars of this type have indeed been found, and are labeled with spectral types L and T.

4.3 Supernovae and Neutron Stars

4.3.1 Core Collapse in Massive Stars

We now return to stars with initial masses (i.e., their masses when they begin their lives on the main sequence) of about $8 M_\odot$ or more. This corresponds to spectral types O and "early" B. After exhausting most of the hydrogen in their cores, such stars move to the giant branch. They then begin a sequence of steps, each consisting of the contraction and heating of the inner regions, resulting in the ignition of new nuclear reactions. As time advances, shells at various inner radii attain the temperatures and the densities required for the reactions that produce progressively heavier elements. Apart from the reactions already discussed for lower-mass stars,

$$^{4}\text{He} + {}^{12}\text{C} \rightarrow {}^{16}\text{O} + \gamma \tag{4.63}$$

and

$$^{4}\text{He} + {}^{16}\text{O} \rightarrow {}^{20}\text{Ne} + \gamma, \tag{4.64}$$

these massive stars can also burn carbon via the reactions

$$^{12}\text{C} + {}^{12}\text{C} \rightarrow {}^{20}\text{Ne} + {}^{4}\text{He} + \gamma, \tag{4.65}$$
$$^{12}\text{C} + {}^{12}\text{C} \rightarrow {}^{23}\text{Na} + p, \tag{4.66}$$

and

$$^{12}\text{C} + {}^{12}\text{C} \rightarrow {}^{23}\text{Mg} + n. \tag{4.67}$$

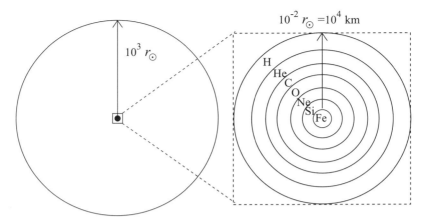

Figure 4.6 Simplified schematic view of the layered structure of a massive star and the distribution of the main elements that compose it, at the onset of core collapse and the ensuing supernova explosion.

Carbon burning is followed by neon, oxygen, and silicon burning. Each of these stages takes less and less time. For example, for a $25 M_\odot$ star, the duration of each burning stage is

H $\sim 5 \times 10^6$ yr
He $\sim 5 \times 10^5$ yr
C ~ 500 yr
Ne ~ 1 yr
Si ~ 1 day.

Massive stars undergo all the stages of nuclear burning up to the production of elements in the "iron group" with atomic mass number around $A = 56$, consisting of isotopes of Cr, Mn, Fe, Co, and Ni. At this stage, the star's outer envelope has expanded to about $1000 r_\odot$, and it has a dense core of radius $\sim 10^4$ km with an onion-like layered structure (see Fig. 4.6). The outer layers of this core are still burning hydrogen. Looking inward, the core consists of concentric shells composed primarily of helium, carbon, oxygen, neon, silicon, and iron, respectively.

Figure 4.7 shows, for all the chemical elements, the binding energy per nucleon (i.e., the binding energy of a nucleus divided by its mass number A). Energy can be gained by fusing or fissioning elements with low binding energy per nucleon into elements with high binding energy per nucleon. The iron group elements are the most tightly bound nuclei, and are therefore a "dead end" in nuclear energy production. Synthesis of iron-group elements into heavier elements consumes, rather than releases, thermal energy. This fact is at the root of the "iron catastrophe" that ensues.

When the central iron core continues to grow and approaches M_{ch}, two processes begin: nuclear photodisintegration and neutronization.

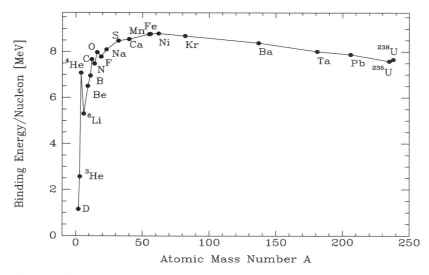

Figure 4.7 Binding energy per nucleon as a function of atomic mass number. Several elements are marked. The iron-group elements with $A \approx 56$ have the highest binding energy per nucleon, 8.8 MeV, and therefore nuclear fusion of these elements into heavier elements does not release thermal energy, but rather consumes it.

Nuclear Photodisintegration: The temperature is high enough for energetic photons to be abundant, and they get absorbed in the endothermic (i.e., energy-consuming) nuclear reaction

$$\gamma + {}^{56}\text{Fe} \rightarrow 13\,{}^{4}\text{He} + 4n, \tag{4.68}$$

with an energy consumption of 124 MeV. The helium nuclei are further unbound in the process

$$\gamma + {}^{4}\text{He} \rightarrow 2p + 2n, \tag{4.69}$$

consuming 28.3 MeV (the binding energy of a ${}^{4}\text{He}$ nucleus). The total energy of the star is thus reduced by $(124 + 13 \times 28.3)/56 \approx 8.8$ MeV$= 1.4 \times 10^{-5}$ erg per nucleon. With about 10^{57} protons in a Chandrasekhar mass, this corresponds to a total energy loss of 1.4×10^{52} erg, ~ 10 times the energy radiated by the Sun over 10^{10} yr.

Neutronization: The large densities in the core lead to a large increase in the rates of processes such as

$$e^- + p \rightarrow n + \nu_e, \tag{4.70}$$

$$e^- + {}^{56}\text{Fe} \rightarrow {}^{56}\text{Mn} + \nu_e, \tag{4.71}$$

$$e^- + {}^{56}\text{Mn} \rightarrow {}^{56}\text{Cr} + \nu_e. \tag{4.72}$$

This neutronization depletes the core of electrons, and their supporting degeneracy pressure, as well as of energy, which is carried off by the neutrinos.

The two processes lead, in principle, to an almost total loss of thermal pressure support and to an unrestrained collapse of the core of a star on a free-fall timescale. For the typical core densities prior to collapse, $\rho \sim 10^9$ g cm^{-3} (calculated from stellar evolution models), this timescale is (Eq. 3.15)

$$\tau_{ff} = \left(\frac{3\pi}{32G\bar{\rho}} \right)^{1/2} \sim 0.1 \text{ s.} \tag{4.73}$$

In practice, at these high densities, the mean free path for neutrino scattering becomes of order the core radius. This slows down the energy loss, and hence the collapse time, to a few seconds.

As the collapse proceeds and the density and the temperature increase, the reaction

$$e^- + p \to n + \nu_e \tag{4.74}$$

becomes common, and is infrequently offset by the inverse process of neutron decay

$$n \to p + e^- + \bar{\nu}_e, \tag{4.75}$$

leading to an equilibrium ratio of densities of

$$n_e = n_p \approx \tfrac{1}{200} n_n. \tag{4.76}$$

Thus, most of the nucleons become neutrons, and a **neutron star** forms, in which degenerate neutrons, rather than electrons, provide the pressure support against gravity.

4.3.2 Properties of Neutron Stars

The properties of neutron stars can be estimated easily by replacing m_e with m_n in Eqs. 4.34 – 4.35, describing white dwarfs. Thus,

$$r_{ns} \approx 2.3 \times 10^9 \text{ cm } \frac{m_e}{m_n} \left(\frac{Z}{A} \right)^{5/3} \left(\frac{M}{M_\odot} \right)^{-1/3} \approx 14 \text{ km} \left(\frac{M}{1.4M_\odot} \right)^{-1/3}. \tag{4.77}$$

Here we have set $Z/A = 1$, since the number of particles contributing to the degeneracy pressure (i.e., the neutrons) is almost equal to the total number of nucleons. Since the radius of a neutron star is about 500 times smaller than that of a white dwarf, the mean density is about 10^8 times greater, i.e., $\rho \sim 10^{14}$ g cm^{-3}. This is similar to the density of nuclear matter. In fact, one can consider a neutron star to be one huge nucleus of atomic mass number $A \sim 10^{57}$.

Our estimate of the radius is only approximate, since we have neglected two effects which are important. First, at these interparticle separations, the nuclear interactions play an important role in the equation of state, apart from the neutron degeneracy pressure. The equation of state of nuclear matter is poorly known, due to our poor understanding of the details of the strong interaction. In fact, it is hoped that actual measurements of the sizes of neutron stars will provide experimental constraints on the nuclear equation of state, which would be important input to nuclear physics. Second, the gravitational potential energy of a test particle of mass m at the surface of a $\sim 1.4 M_\odot$ neutron star, of radius $r \sim 10$ km, is a significant fraction of the particle's rest-mass energy:

$$\frac{E_{gr}}{mc^2} = \frac{GM}{rc^2} \approx \frac{6.7 \times 10^{-8} \text{ cgs} \times 1.4 \times 2 \times 10^{33} \text{ g}}{10 \times 10^5 \text{ cm } (3 \times 10^{10} \text{ cm s}^{-1})^2} \approx 20\%. \tag{4.78}$$

This means that matter falling onto a neutron star loses 20% of its rest mass, and the mass of the star (as measured, e.g., via Kepler's law) is 20% smaller than the total mass that composed it. Thus, a proper treatment of the physics of neutron stars needs to be calculated within the strong-field regime of general relativity. More detailed calculations, including these two effects, give a radius of about 10 km for a $1.4 M_\odot$ neutron star.

The Chandrasekhar mass,

$$M_{ch} = 0.2 \left(\frac{\mathcal{Z}}{A} \right)^2 \left(\frac{hc}{Gm_p^2} \right)^{3/2} m_p, \tag{4.79}$$

can be used to estimate a maximal mass for a neutron star, beyond which the density is so high that even the degenerate neutron gas becomes ultra-relativistic and unable to support the star against gravity. Again replacing the $\mathcal{Z}/A = 0.5$, appropriate for white dwarfs, with $\mathcal{Z}/A = 1$, describing neutron stars, gives a factor of 4, or

$$M_{ns,max} = 1.4 M_\odot \times 4 = 5.6 M_\odot. \tag{4.80}$$

Taking into account general relativistic effects lowers this estimate to about $5 M_\odot$. This reduction come about because, in the regime of strong gravity, the pressure itself contributes significantly to the gravitational field, and thus pressure gradually loses its effectiveness in counteracting gravitation. Detailed calculations that attempt also to take into account the strong interaction of nuclear matter further lower the maximal mass to 2–$3 M_\odot$, but this is still highly uncertain.

4.3.3 Supernova Explosions

The fall of the layers of matter that surrounded the core onto the surface of a newly formed neutron star sets off a shock wave that propagates outward and blows off the outer shells of the star, in what is observed as a **supernova** explosion (see Fig. 4.8). The details of how exactly this occurs are not understood yet. In fact, sophisticated numerical simulations of the collapse are presently still unable to reproduce all the properties of the observed "explosion," i.e., the ejection of the star's outer regions. A kinetic energy of about 3×10^{51} erg is imparted to the material flying out (as determined from measurements of the mass and velocity of ejecta in supernova remnants). About 3×10^{49} erg can be observed over a period of order one month as luminous energy, driven primarily by the decay of radioactive elements synthesized during the last few moments before collapse, during the collapse, and during the explosion. Although the luminous energy is only 1% of the kinetic energy, it nevertheless makes a supernova an impressive event; the mean luminosity is of order

Figure 4.8 Optical-light image of the supernova SN1994D, below and to the left of its host galaxy, NGC 4526. For several weeks around its peak brightness, the luminosity of a supernova is comparable to that of an entire galaxy, with $L \sim 10^{10} L_\odot$. (The spikes emerging from the supernova are due to diffraction.) Photo credit: NASA, ESA, the High-Z Supernova Search Team and the Hubble Key Project Team.

$$L_{SN} \sim \frac{3 \times 10^{49} \text{ erg}}{30 \text{ d} \times 24 \text{ hr} \times 60 \text{ m} \times 60 \text{ s}} \sim 10^{43} \text{ erg s}^{-1} \sim 3 \times 10^9 L_\odot, \tag{4.81}$$

comparable to the luminosity of an entire galaxy of stars (see chapter 6).

However, the total gravitational binding energy released in the collapse of the core to a neutron star is

$$E_{gr} \sim \frac{GM^2}{r_{ns}} = 5 \times 10^{53} \left(\frac{M}{1.4 M_\odot}\right)^2 \left(\frac{r_{ns}}{10 \text{ km}}\right)^{-1} \text{ erg}. \tag{4.82}$$

The kinetic and radiative energies are just small fractions, $\sim 10^{-2}$ and $\sim 10^{-4}$, respectively, of this energy. The bulk of the energy released in the collapse is carried away by neutrino–antineutrino pairs. The density is so high that photons cannot emerge from the star, and they undergo frequent photon–photon collisions. These produce electron–positron pairs, which form neutrino pairs:

$$\gamma + \gamma \rightarrow e^+ + e^- \rightarrow \nu_e + \bar{\nu}_e, \nu_\mu + \bar{\nu}_\mu, \nu_\tau + \bar{\nu}_\tau \tag{4.83}$$

(the μ and τ neutrinos are neutrinos related to the muon and the tauon, which are heavy relatives of the electron.) The neutrinos can pass through the star with few scatterings (see Problem 3), and can therefore drain almost all of the thermal energy.

A striking confirmation of this picture was obtained in 1987, with the explosion of Supernova 1987A in the Large Magellanic Cloud, a satellite galaxy of our Galaxy (the Milky Way; see chapter 6), at a distance of 50 kpc from Earth. This was the nearest supernova observed since the year 1604. A total of 20 antineutrinos (several of them with directional information pointing toward the supernova) were detected simultaneously in the span of a few seconds by two different underground experiments. Each experiment consisted of a detector composed of a large tank filled with water and surrounded by photomultiplier tubes. These experiments were initially designed to search for proton decay. The experiments discovered the antineutrinos, and measured their approximate energies and directions via the reaction

$$\bar{\nu}_e + p \rightarrow n + e^+, \tag{4.84}$$

by detecting the Cerenkov radiation emitted by the positrons moving faster than the speed of light in water. The typical energies of the $\bar{\nu}_e$'s were 20 MeV. The detection of 20 particles, divided by the efficiency of the experiments to antineutrino detection (which was a function of antineutrino energy), implied that a *fluence* (i.e., a time-integrated flux) of 2×10^9 cm^{-2} electron antineutrinos had reached Earth. The electron antineutrinos, $\bar{\nu}_e$'s, are just one out of six types of particles ($\nu_e, \bar{\nu}_e, \nu_\mu, \bar{\nu}_\mu, \nu_\tau, \bar{\nu}_\tau$) that are produced in similar numbers and carry off the collapse energy. Thus, the total energy released in neutrinos was

$$\begin{aligned} E_{\text{neutrino}} &\sim 2 \times 10^9 \text{ cm}^{-2} \times 6 \times (20 \text{ MeV} \times 1.6 \times 10^{-6} \text{erg MeV}^{-1}) \\ &\times 4\pi (5 \times 10^4 \text{ pc} \times 3.1 \times 10^{18} \text{ cm pc}^{-1})^2 \sim 10^{53} \text{ erg}, \end{aligned} \tag{4.85}$$

close to the total energy expected from the collapse of a stellar core.

We note here that there is an altogether different avenue for stars to pass the Chandrasekhar limit and explode, in events that are called **type Ia supernovae**. White dwarfs that are in close binaries, where mass transfer takes place from a companion onto the white dwarf, can reach M_{ch}. At that stage, or possibly even before actually reaching M_{ch}, thermonuclear ignition of the carbon core occurs. However, this happens under degenerate conditions, without the thermostatic effect of a classical equation of state. With a classical equation of state, a rise in temperature produces a rise in pressure that leads to an expansion of the star, a lowering of the temperature, and a decrease in the nuclear reaction rates. Instead, under degenerate conditions, the white dwarf structure is insensitive to the rise in temperature, which raises the nuclear reaction rates more and more, ending in a thermonuclear runaway that blows up the entire star. As opposed to core-collapse supernovae, which leave a neutron star remnant (or a black hole, see below), it is believed that type Ia supernovae leave no stellar remnant (see Problem 4). It is presently unknown what kind of star is the companion of the white dwarf in the systems that are the progenitors of of type Ia supernovae. It is also possible that a type-Ia explosion is actually the result of the merger of two white dwarfs—see Problem 6. The supernova shown in Fig. 4.8 was a type Ia event. We will return later to the physics of accretion in close binaries (section 4.6) and to the use of type Ia supernovae in cosmology (chapters 7 and 9).

Finally, there exists a class of objects even more luminous than SNe, though very transient, called **gamma-ray bursts** (GRBs). These explosions release of the order of 10^{51} erg over a period of just a few seconds. As their name implies, much of this energy is released at gamma-ray frequencies, but the rapidly fading "afterglows" of the explosions can sometimes be detected at longer wavelengths on longer timescales—minutes in X-rays, days in the optical, and weeks at radio wavelengths. A GRB occurs about once a day in the observable Universe. It is now known that at least half of these explosions occur in star-forming galaxies, i.e., galaxies that have massive young stars. This argues that some GRBs result from the core collapse of massive stars of a particular type or in a particular configuration, i.e., that they are a certain kind of core-collapse supernova. In recent years, evidence has been accumulating that actually links some GRBs to supernovae observed at the same location. The nature and mechanisms of GRBs are still widely debated. The large energy outputs, as well as indirect evidence of the existence of highly relativistic bulk motions of material, suggest that GRBs involve the formation of black holes.

The material expelled by the mass outflows from giants and by both types of supernovae—core-collapse and Ia—is essentially the only source of all "heavy" elements. Except for helium and trace amounts of the next few lightest elements, which were synthesized early in the history of the Universe (as we will see in chapter 9), all *nucleosynthesis* takes place inside stars during their various evolution stages, or during their explosions as supernovae.

4.4 Pulsars and Supernova Remnants

Many neutron stars have been identified as such in their manifestation as **pulsars**. Pulsars were first discovered with radio telescopes in the 1960s as point sources of periodic pulses of radio emission, with periods of the order of $\tau \sim 10^{-3}$ to 1 s. Today, over 1000 pulsars are known. The periods of most pulsars are observed to grow slowly with time in a very regular manner. The predictability of the pulse arrival times is comparable to that of the most accurate man-made clocks. Figure 4.9 shows a typical pulsar time series. One of the best studied pulsars, which we shall use as an example, is the Crab pulsar, at the center of the Crab nebula (see Fig. 4.10). The Crab nebula, an example of a **supernova remnant**, is an expanding cloud of gaseous fragments at the same location in the sky where a bright supernova explosion was observed and documented in the year 1054 by Chinese, Japanese, and Korean astronomers. The Crab pulsar, from which pulsations are detected at radio, optical, and X-ray wavelengths, has a pulsation period of $\tau = 33$ ms, i.e., an angular frequency

$$\omega = \frac{2\pi}{\tau} = 190 \text{ s}^{-1}. \tag{4.86}$$

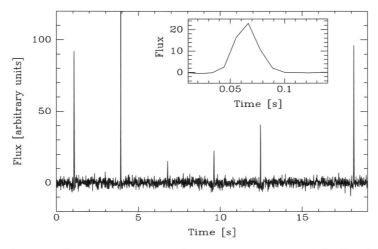

Figure 4.9 Flux at 430 MHz vs. time from PSR J0546+2441, a typical radio pulsar, over several periods. The pulse period is 2.84385038524 s (i.e., measured to 12 significant digits). Note the variable strength, and occasional disappearance of the pulses. The inset shows a zoom-in on the pulse profile, averaged over many periods. Data credit: D. Champion, see *Mon. Not. Royal. Astron. Soc.* (2005), 363, 929.

The period derivative is

$$\frac{d\tau}{dt} = \frac{1 \text{ ms}}{75 \text{ yr}} = 4.2 \times 10^{-13}, \tag{4.87}$$

or

$$\frac{d\omega}{dt} = -\frac{2\pi}{\tau^2}\frac{d\tau}{dt} = -2.4 \times 10^{-9} \text{ s}^{-2}. \tag{4.88}$$

The total luminosity of the Crab nebula, integrated over all wavelengths, is

$$L_{\text{tot}} \approx 5 \times 10^{38} \text{erg s}^{-1}, \tag{4.89}$$

and is mostly in the form of **synchrotron radiation**, i.e., radiation emitted by relativistic electrons as they spiral along magnetic field lines.

4.4.1 Identification of Pulsars as Neutron Stars

To see that the Crab pulsar (and other pulsars) are most plausibly identified with spinning neutron stars, let us consider possible mechanisms for producing periodicity of the observed magnitude and regularity. Three options that come to mind, of astronomical phenomena associated with periodicity, are binaries, stellar pulsations, and stellar rotation. For binary orbits, the angular frequency, masses, and separation are related by Kepler's law,

Figure 4.10 The Crab nebula, the remnant of a core-collapse supernova that exploded in the year 1054, at a distance of 2 kpc. *Top:* Image in optical light. Image scale is 4 pc on a side. *Bottom:* Zoom in on the area marked in the top photo, in optical light (left), with the pulsar at the center of the remnant indicated by the arrow; and in X-rays (right), showing the pulsar, bidirectional jets, and a toroidal structure formed by synchrotron emission from energetic particles. Note that a similar emission morphology is faintly discernible also in the optical image on the left. Photo credits: European Southern Observatory; and NASA/CXC/ASU and J. Hester et al.

$$\omega^2 = \frac{G(M_1 + M_2)}{a^3}, \tag{4.90}$$

or

$$\begin{aligned}
a &= \frac{[G(M_1 + M_2)]^{1/3}}{\omega^{2/3}} \\
&= \frac{[6.7 \times 10^{-8} \text{ cgs } (4 \times 10^{33} \text{ g})]^{1/3}}{(190 \text{ s}^{-1})^{2/3}} = 2 \times 10^7 \text{ cm} = 200 \text{ km}, \tag{4.91}
\end{aligned}$$

where we have assumed two solar-mass objects and inserted the Crab pulsar's frequency. The separation a is much smaller than the radii of normal stars or of white dwarfs. Only a pair of neutron stars could exist in a binary at this separation. However, general relativity predicts that two such masses orbiting at so small a separation will lose gravitational binding energy via the emission of gravitational waves (see Problems 5 and 6). This loss of energy will cause the separation between the pair to shrink, and the orbital frequency to *grow*, contrary to the observation that the pulsar frequencies decrease with time. Thus, orbital motion of stellar-mass objects cannot be the explanation for pulsars.

A second option is stellar pulsations. Stars are, in fact, observed to pulsate regularly in various modes, with the pulsation period dependent on density as[6] $\tau \propto \rho^{-1/2}$. Normal stars oscillate with periods between hours and months, and white dwarfs oscillate with periods of 100 to 1000 s. Neutron stars, which are 10^8 times denser than white dwarfs, should therefore pulsate with periods 10^4 times shorter, i.e., less than 0.1 s. However, the most common period for pulsars is about 0.8 s. There is thus no known class of stars with the density that would produce the required pulsation period.

Finally, let us assume that the rapid and very regular pulsation is produced via anisotropic emission from a rotating star. The fastest that a star can spin is at the angular frequency at which centrifugal forces do not break it apart. This limit can be found by requiring that the gravitational force on a test mass m, at the surface, be greater than the centrifugal force:

$$\frac{GMm}{r^2} > m\omega^2 r, \tag{4.92}$$

or

$$\frac{M}{r^3} > \frac{\omega^2}{G}, \tag{4.93}$$

and therefore

$$\bar{\rho} = \frac{3M}{4\pi r^3} > \frac{3\omega^2}{4\pi G} = \frac{3(190 \text{ s}^{-1})^2}{4\pi \times 6.7 \times 10^{-8} \text{ cgs}} = 1.3 \times 10^{11} \text{ g cm}^{-3}, \tag{4.94}$$

for the Crab pulsar. Thus, if the Crab pulsar is a spinning star and is not flying apart, its mean density must be at least five orders of magnitude larger than that of a white dwarf,

---

[6] It is easy to see from a dimensional argument that this must be the case for radial pulsations. Consider a star that is "squeezed" radially, and then released. The restoring force due to the pressure has dimensions of pressure times area, $F \sim PA \sim (GM\rho/r)r^2$, where we have used the equation of hydrostatic equilibrium (Eq. 3.19) to express the dimensions of the pressure. Equating this to the mass times the acceleration, $Ma \sim Mr/\tau^2$, gives the required result. Note that the pulsation period, $\tau \sim (G\rho)^{-1/2}$, is essentially the same as the free-fall timescale, Eq. 3.15.

but consistent with that of a neutron star. Note that the pulsars with the shortest periods known, of about 1 ms (rather than the Crab's 33 ms), must have mean densities 1000 times larger to avoid breaking apart, i.e., $\sim 10^{14}$ g cm^{-3}. This is just the mean density we predicted for neutron stars.

Next, let us presume that the luminosity of the Crab nebula is powered by the pulsar's loss of rotational energy as it spins down. (The observed luminosity of the pulsar itself, $\sim 10^{31}$ erg s^{-1}, is much too small to be the energy source of the extended emission.) Since

$$E_{\rm rot} = \tfrac{1}{2} I \omega^2, \tag{4.95}$$

where I is the moment of inertia,

$$L_{\rm tot} = -\frac{dE_{\rm rot}}{dt} = -I\omega \frac{d\omega}{dt}. \tag{4.96}$$

For an order-of-magnitude estimate, let us use the moment of inertia of a constant-density sphere, $I = \tfrac{2}{5} M r^2$. Then

$$Mr^2 \sim -\frac{5}{2} \frac{L_{\rm tot}}{\omega d\omega/dt} = -\frac{5 \times 5 \times 10^{38} \text{ erg s}^{-1}}{2 \times 190 \text{ s}^{-1}(-2.4 \times 10^{-9} \text{ s}^{-2})} = 3 \times 10^{45} \text{ g cm}^2. \tag{4.97}$$

A $1.4 M_\odot$ neutron star of radius 10 km has just this value of Mr^2:

$$Mr^2 = 1.4 \times 2 \times 10^{33} \text{ g} \times (10^6 \text{ cm})^2 = 2.8 \times 10^{45} \text{ g cm}^2. \tag{4.98}$$

By comparison, a normal star like the Sun has Mr^2 of order 10^9 larger than the value in Eq. 4.98. Conservation of angular momentum, $J = I\omega$, then dictates that when a stellar core of solar mass and solar radius collapses to a radius of about 10 km, it will spin up by a factor of order 10^9. The rotation period of the Sun is 25.4 days, or 2×10^6 s, which is typical of main-sequence stars. Collapse of a stellar core to neutron-star proportions is thus expected to produce an object with a spin period of order milliseconds, as observed in pulsars.

Thus, we see that if we identify pulsars as rapidly spinning stars, then their spin rate is that expected from the collapse of the cores of main-sequence stars to neutron star dimensions; their mean densities are those of neutron stars; and their loss or rotational energy accounts for the luminosity of the supernova ejecta in which they are embedded, if they have the moments of inertia of neutron stars. Finally, the location of pulsars at the sites of some historical supernovae, an explosion that is expected to accompany the formation of a neutron star (in terms of the energy released, even if the details of the explosion are not yet fully understood), leaves little doubt that pulsars are indeed neutron stars.

4.4.2 Pulsar Emission Mechanisms

The details of how pulsars produce their observed periodic emission are still a matter of active research. However, it is widely accepted that the basic phenomenon is the rotation of a neutron star having a magnetic field axis that is misaligned with respect to the star's rotation axis by some angle θ. A spinning magnetic dipole radiates an electromagnetic luminosity

$$L = \frac{1}{6c^3} B^2 r^6 \omega^4 \sin^2 \theta, \tag{4.99}$$

where B is the magnetic field on the surface of the star, at a radius r, on the magnetic pole. Solving Eq. 4.99 for B, with the observed properties of the Crab, a typical neutron-star radius, and $\sin \theta \approx 1$,

$$B \sim \frac{(6c^3 L)^{1/2}}{r^3 \omega^2 \sin \theta} \sim \frac{[6(3 \times 10^{10} \text{ cm s}^{-1})^3 \times 5 \times 10^{38} \text{ erg s}^{-1}]^{1/2}}{(10^6 \text{ cm})^3 (190 \text{ s}^{-1})^2 \times 1} \sim 8 \times 10^{12} \text{ G.} \tag{4.100}$$

Magnetic fields of roughly such an order of magnitude are expected when the ionized (and hence highly conductive) gas in a star is compressed during the collapse of the iron core. The originally small magnetic field of the star (e.g., ~ 1 G in the Sun) is "frozen" into the gas. When the gas is compressed, the flux in the magnetic field lines is amplified in proportion to r^{-2}, corresponding to $\sim 10^{10}$ between the core of a main sequence star and a neutron star.

In a process that is not yet fully agreed upon, the complex interactions between magnetic and electric fields, particles, and radiation in the neighborhood of the neutron star power the nebula, and also lead to the emission of radiation in two conical beams in the direction of the magnetic axis. As the star spins and the magnetic axis precesses around the rotation axis, each beams traces an annulus of angular radius θ on the sky, as seen from the neutron star (see Fig. 4.11). Distant observers who happen to lie on the path of these "lighthouse beams" detect a pulse once every rotation, when the beam sweeps past them. This implies, of course, that we can detect only a fraction of all pulsars, namely those for which the Earth lies in the path of one of the beams.

Evidence that magnetic dipole radiation is the basic emission mechanism can be found from the age of the Crab pulsar. If such radiation is leading to the pulsar's loss of rotational energy, then, combining Eqs. 4.96 and 4.99, we find

$$\frac{dE_{\text{rot}}}{dt} = I\omega \frac{d\omega}{dt} \propto \omega^4, \tag{4.101}$$

and

$$\frac{d\omega}{dt} = C\omega^3. \tag{4.102}$$

The constant C can be determined from the present values of $d\omega/dt$ and ω,

$$C = \frac{\dot{\omega}_0}{\omega_0^3}. \tag{4.103}$$

Separating variables in Eq. 4.102 and integrating, we obtain for the age of the pulsar

$$t_{\text{pulsar}} = \frac{\omega_0^3}{2\dot{\omega}_0} \left(\frac{1}{\omega^2} - \frac{1}{\omega_i^2} \right), \tag{4.104}$$

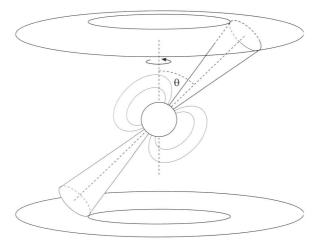

Figure 4.11 Schematic model of a pulsar. Biconical beams of radiation emerge along the magnetic axis of a neutron star. The magnetic axis is inclined by an angle θ to the star's spin axis. Observers in a direction, as viewed from the star, that is within one of the two annular regions swept out by the beams as the star rotates will detect periodic pulses.

where ω_i is the initial angular frequency of the neutron star upon formation. Thus, an upper limit on the current age of the Crab pulsar is obtained by taking $\omega = \omega_0$ and $\omega_i = \infty$,

$$t_{\text{pulsar}} < \frac{\omega_0}{2\dot{\omega}_0} = \frac{190 \text{ s}^{-1}}{2 \times 2.4 \times 10^{-9} \text{ s}^{-2}} = 4 \times 10^{10} \text{ s} = 1260 \text{ yr}. \tag{4.105}$$

This limit is consistent with the historical age, 950 yr, of the supernova of the year 1054. The pulsar age will *equal* 950 yr if we set $\tau_i = 2\pi/\omega_i = 2.5$ ms, close to the expected spin rate of newly formed neutron stars.[7]

4.4.3 Neutron Star Cooling

As already noted, according to the above picture, we observe only a fraction of all pulsars, those for which the Earth is in the rotating pulsar beam. More significantly, pulsars slow down and lose their rotational energies, and as a result, at some point in time, will become undetectable as pulsars. However, there should exist a large population of old, spun-down, neutron stars—the remnants of all massive stars that have undergone core collapse to this state. In section 4.2.3.3, we saw that the small surface areas of white dwarfs result in very long cooling times. The surfaces of neutron stars, smaller by five orders of magnitude compared to those of white dwarfs, mean that old neutron stars will be "stuck" at temperatures of order 10^5 K, with thermal radiation peaking at photon energies of tens of electron volts (called the *extreme UV* range).

---

[7] The so-called *braking index*, which equals 3 in Eq. 4.102 for the case of a magnetic dipole, has actually been measured directly for several pulsars, and is sometimes less than 3. Such is the case for the Crab pulsar, and its deduced initial spin period is then actually 19 ms.

Detailed calculations of neutron star cooling are considerably more uncertain than those for white dwarfs, partly due to the poorly constrained equation of state on nuclear matter, which leads to uncertainty in the structure and composition of a neutron star. A cooling calculation also needs to take into account many different physical processes, not all fully understood, that may play a role under the extreme conditions of gravity, temperature, density, and magnetic field inside and near the surface of a neutron star. Interstellar gas atoms falling onto a neutron-star surface also have an effect, and are likely to heat it to X-ray temperatures. To date, only several candidate isolated old neutron stars have been found in X-ray surveys. The small surface areas of neutron stars mean that their optical luminosities are very low, and hence such objects can be found only when they are near enough. X-ray surveys do reveal a large population of accreting neutron stars in binary systems, called *X-ray binaries*, which we will study in section 4.6.

4.5 Black Holes

In the case of a stellar remnant with a mass above the maximum allowed mass of a neutron star, no mechanism is known that can prevent the complete gravitational collapse of the object. In fact, general relativity predicts that even if some new form of pressure sets in at high densities, the gravitational field due to such pressure will overcome any support the pressure gradient provides, and the collapse of the star to a singularity, or **black hole** is unavoidable.

As its name implies, matter or radiation cannot escape from a black hole. An incorrect derivation, giving the correct answer, of the degree to which a mass must be compressed to become a black hole can be obtained by requiring that the escape velocity, v_e, from a spherical mass of radius r be greater than c (and hence nothing can escape),

$$\frac{GM}{r} > \frac{1}{2}v_e^2 = \frac{1}{2}c^2, \tag{4.106}$$

and therefore the **Schwarzschild radius** is

$$r_s = \frac{2GM}{c^2} = 3 \text{ km} \frac{M}{M_\odot}. \tag{4.107}$$

Photons cannot escape from an object with a mass M that is concentrated within a radius smaller than r_s. The above derivation is incorrect because the kinetic energy of a photon is not $mc^2/2$, nor is the gravitational potential accurately described by the Newtonian limit, GM/r.

A correct derivation of r_s, which we shall only outline schematically, begins with the Einstein equations of general relativity,

$$G_{\mu\nu} = \frac{8\pi G}{c^4} T_{\mu\nu}. \tag{4.108}$$

The Einstein equations relate the geometry and curvature of spacetime to the distribution of mass–energy. $T_{\mu\nu}$ is the *energy–momentum tensor*. It is represented by a 4×4 matrix, and each of its indices runs over the four spacetime coordinates. This is the "source" term in the equations and includes mass–energy density and pressure. $G_{\mu\nu}$ is the *Einstein tensor* consisting of combinations of first and second partial derivatives, with respect to the spacetime coordinates, of the **metric**, $g_{\mu\nu}$. (A more detailed description of $T_{\mu\nu}$ and $G_{\mu\nu}$ is given in chapter 8.) The metric describes the geometry of spacetime via the *line element*

$$(ds)^2 = \sum_{\mu,\nu} g_{\mu\nu} dx_\mu dx_\nu, \tag{4.109}$$

where ds is the interval between two close spacetime events. For example, the metric (familiar from special relativity) that describes spacetime in a flat (*Euclidean*) region of space, far from any mass concentration, is the **Minkowski metric**, with a line element

$$(ds)^2 = (cdt)^2 - (dx)^2 - (dy)^2 - (dz)^2. \tag{4.110}$$

The nonzero terms of the 4×4 matrix describing $g_{\mu\nu}$ in this case are

$$g_{00} = 1, \quad g_{11} = -1, \quad g_{22} = -1, \quad g_{33} = -1. \tag{4.111}$$

In spherical coordinates, the Minkowski metric has the form,

$$(ds)^2 = (cdt)^2 - (dr)^2 - (rd\theta)^2 - (r \sin\theta d\phi)^2, \tag{4.112}$$

i.e.,

$$g_{00} = 1, \quad g_{11} = -1, \quad g_{22} = -r^2, \quad g_{33} = -r^2 \sin^2\theta. \tag{4.113}$$

Since $G_{\mu\nu}$ and $T_{\mu\nu}$ are symmetric 4×4 tensors (e.g., $G_{\mu\nu} = G_{\nu\mu}$), there are only 10, rather than 16, independent Einstein equations, and a zero-divergence condition on $T_{\mu\nu}$ (implying local energy conservation) further reduces this to six equations.

A solution of the Einstein equations for the geometry of spacetime in the vacuum surrounding a static, spherically symmetric, mass distribution, as viewed by an observer at infinity (i.e., very distant from the mass) is the **Schwarzschild metric:**

$$(ds)^2 = \left(1 - \frac{2GM}{rc^2}\right)(cdt)^2 - \left(1 - \frac{2GM}{rc^2}\right)^{-1}(dr)^2 - (rd\theta)^2 - (r \sin\theta d\phi)^2, \tag{4.114}$$

where r, θ, and ϕ are spherical coordinates centered on the mass, and t is the time measured by the distant observer. The time shown by any clock can be found from the **proper time** τ, defined as

$$d\tau \equiv \frac{ds}{c}. \tag{4.115}$$

For a clock at rest (i.e., $dr = d\theta = d\phi = 0$),

$$d\tau = \left(1 - \frac{2GM}{rc^2}\right)^{1/2} dt = \left(1 - \frac{r_s}{r}\right)^{1/2} dt. \tag{4.116}$$

Consider now a stellar remnant that is compact enough that its radius is within r_s, and hence the Schwarzschild metric (which applies only in vacuum) describes spacetime in the vicinity of r_s. When a clock is placed at $r \rightarrow r_s$, $d\tau$ approaches zero times dt. During a time interval of, say, $dt = 1$ s, measured by a distant observer, the clock near r_s advances by much less. In other words, clocks appear (to a distant observer) to tick more and more slowly as they approach r_s, and to stop completely at r_s. This is called **gravitational time dilation**.

The electric and magnetic fields of a light wave emitted by a source near r_s will also appear to oscillate more slowly due to the time dilation, and therefore the frequency of light will decrease, and its wavelength λ will increase, relative to the wavelength λ_0 of light emitted by the same source far from the black hole. This **gravitational redshift** is

$$\frac{\lambda}{\lambda_0} = \left(1 - \frac{2GM}{rc^2}\right)^{-1/2} = \left(1 - \frac{r_s}{r}\right)^{-1/2}. \tag{4.117}$$

When the light source is at r_s, the wavelength becomes infinite and the energy of the photons, hc/λ, approaches zero.

In general relativity, once we know the metric that describes spacetime, we can find the trajectories of free-falling particles and of radiation. In particular, massless particles and light move along **null geodesics**, defined as paths along which $ds = 0$. Setting $ds = 0$ in Eq. 4.114, the *coordinate speed* of a light beam moving in the radial direction $(d\theta = d\phi = 0)$ is

$$\frac{dr}{dt} = \pm c \left(1 - \frac{2GM}{rc^2}\right) = \pm c \left(1 - \frac{r_s}{r}\right). \tag{4.118}$$

At $r \gg r_s$, the speed is $\pm c$, as expected. However, as light is emitted from closer and closer to r_s, its speed appears to decline (again, to a distant observer), going to zero at r_s. Gravity works effectively as an index of refraction, with $n = \infty$ at r_s. As a result, no information can emerge from a radius smaller than r_s, which constitutes an **event horizon** around the black hole. We have thus rederived (correctly, this time) the Schwarzschild radius and its main properties.

Because of gravitational time dilation, a star collapsing to a black hole, as viewed by a distant observer, appears to shrink in progressively slower motion, and gradually appears to "freeze" as it approaches its Schwarzschild radius. In fact, it takes an infinite time for the star to cross r_s, and therefore, formally, black holes do not exist, in terms of distant static observers such as ourselves. (They certainly can exist, even in the "present" of observers who are near enough to a black hole.) However, for all practical purposes, there are no differences in observed properties between such "frozen stars" and truly collapsed black holes. This comes about because, as a source of light falls toward r_s, the rate at which photons from the source reach the observer declines as $(1 - r_s/r)^{1/2}$. Furthermore, the energy of each photon declines due to the gravitational redshift also as $(1 - r_s/r)^{1/2}$. The equation of motion for a radially free-falling light source, $r(t)$, can be roughly estimated by noting that, as the source approaches r_s, it will move with a velocity close to c, and hence its geodesic (i.e., its path in spacetime) will be close to that of the null geodesic of photons. Let us take the negative solution in Eq. 4.118 (the source is falling to smaller radii). Separating the variables,

$$cdt = -\frac{rdr}{r - r_s},$$ (4.119)

changing variables to $x = r - r_s$, and integrating gives

$$c(t - t_0) = -\int \frac{x + r_s}{x} dx = -(x + r_s \ln x) = -[r - r_s + r_s \ln(r - r_s)].$$ (4.120)

As $r - r_s \to 0$, the logarithmic term becomes dominant, and we can write the equation of motion, $r(t)$, as viewed by the distant observer, as

$$r - r_s \sim \exp\left[-\frac{c(t - t_0)}{r_s}\right].$$ (4.121)

Inserting this dependence into the expression for the decline in the observed photon emission rate due to gravitational time dilation, we find that

$$\frac{dN_{ph}}{dt} \sim \left(1 - \frac{r_s}{r}\right)^{1/2} = \left(\frac{r - r_s}{r}\right)^{1/2} \propto \exp\left(-\frac{ct}{2r_s}\right)$$ (4.122)

(where in the last step we have substituted $r \sim r_s$ in the denominator, and $r - r_s$ in the numerator using Eq. 4.121).

For example, for a $5 M_\odot$ stellar core undergoing its final collapse, the characteristic time is

$$\frac{2r_s}{c} = \frac{2 \times 5 \times 3 \text{ km} \times 10^5 \text{ cm km}^{-1}}{3 \times 10^{10} \text{ cm s}^{-1}} = 10^{-4} \text{ s} = 0.1 \text{ ms}.$$ (4.123)

Thus, after a mere 20 ms, the photon rate will decline by a factor $\exp(-200) = 10^{-87}$. The photon emission rate from a Sun-like star emitting in the optical range, at a typical photon energy of $h\nu = 1$ eV, is of order

$$\frac{dN_{ph}}{dt} \sim \frac{L}{h\nu} \sim \frac{3.8 \times 10^{33} \text{ erg s}^{-1}}{1 \text{ eV} \times 1.6 \times 10^{-12} \text{ erg eV}^{-1}} \sim 10^{45} \text{ s}^{-1}.$$ (4.124)

A factor of 10^{87} decrease in photon flux implies that, after just 20 ms, the photon emission rate from the star will decrease to $\sim 10^{-42}$ s^{-1}. The time between the emission of consecutive photons will thus be $\sim 10^{42}$ s, many orders of magnitude larger than the age of the Universe, which is of order 10^{10} yr $\sim 10^{17}$ s. The "frozen star" is truly "black," and no photons emerge from it after a timescale of milliseconds.

Theoretically, quantum mechanics allows an exception to this rule, and small amounts of so-called *Hawking radiation* can escape a black hole, even causing it to "evaporate" completely if it is small enough. However, it is unclear if black hole evaporation has any astronomical relevance.

Observationally, there are many objects considered to be stellar-mass black hole candidates, consisting of members of binary systems in which the minimum mass of one of the members is significantly larger than $3 M_\odot$, yet a main-sequence or giant star of such mass is not seen. Presumably, black holes form from the core collapse of stars with an initial mass above some threshold (which is currently thought to be about $25 M_\odot$). In some of these binary systems, accretion of matter onto the black hole is taking place. Such systems are discussed in more detail in section 4.6. Finally, there is evidence for the existence of *supermassive* black holes, with masses of $\sim 10^6 – 10^9 M_\odot$, in the centers of most large galaxies. These are discussed in chapter 6.

4.6 Interacting Binaries

Until now, stars were the only luminous objects we considered. However, there exists an assortment of objects that are powered not by nuclear reactions, but by the accretion of matter onto a gravitational potential well. Objects in this category include pre-main-sequence stars, interacting binaries, active galactic nuclei and quasars, and possibly some types of supernovae and gamma-ray bursts. While all these objects are rare relative to normal stars, they are interesting and important for many physical and observational applications. The physics of accretion is similar in many of these objects. In this section, we will focus on interacting binaries, which are the best-studied accretion powered objects.

As already noted, many stars are in binary systems.[8] Pairs with an orbital period of less than about 10 days are usually in orbits that are circular, "aligned" (i.e., the spin axes of the two stars and the orbital plane axis are all parallel), and synchronized (i.e., each star completes a single rotation about its axis once per orbit, and thus each star always sees the same side of its companion star). This comes about by the action of the strong **tidal forces** that the stars exert on each other at small separations. The force per unit mass on a mass element at the surface of a star, at distance Δr from the center, due to the mass M_1 of the star itself is

$$\frac{F_{\text{grav}}}{m} = \frac{GM_1}{(\Delta r)^2}. \tag{4.125}$$

The tidal force on this mass element, due to the influence of the second star of mass M_2 at a distance r (assuming $\Delta r \ll r$) is

$$\frac{F_{\text{tide}}}{m} = GM_2 \left(\frac{1}{r^2} - \frac{1}{(r + \Delta r)^2} \right) \approx \frac{2GM_2 \Delta r}{r^3}. \tag{4.126}$$

The ratio between the forces is

$$\frac{F_{\text{tide}}}{F_{\text{grav}}} = \frac{2M_2}{M_1} \left(\frac{\Delta r}{r} \right)^3. \tag{4.127}$$

Thus, the larger $\Delta r/r$, the more tidal distortion of the shapes of the stars occurs, such that they become two ovals pointing at each other. As long as the stars are not tidally locked (i.e.,

---

[8] Current evidence is that the binary fraction among stars depends on stellar mass, with most of the massive stars being in binaries, but most low-mass stars being single. About one-half of solar-mass stars are in binaries.

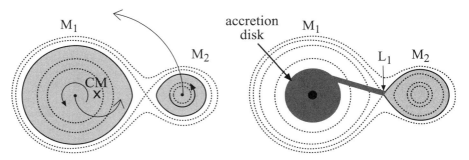

Figure 4.12 Equipotential surfaces (dotted curves) in the corotating reference frame of a binary system with mass ratio $M_1/M_2 = 5$. *Left:* In this example, both stars are inside their respective Roche lobes, but are tidally distorted. Loss of energy to tidal friction will cease only when the orbits about the center of mass become circularized, and aligned and synchronized with the rotations, so that there is no motion in the corotating frame. *Right:* Here, the secondary star, on the right, fills its Roche lobe. Matter flows through the L_1 point and falls onto the primary star, which is now a compact object. Viewed from an inertial frame, the falling material possesses angular momentum, and hence an accretion disk is formed around the compact primary star.

synchronized and circularized), energy is continuously lost to friction while the different parts of each star are deformed during the orbit. Once tidal locking is achieved, everything appears stationary in a reference frame rotating at the binary frequency, and the system achieves its minimum energy.[9]

If we draw the surfaces of constant potential energy in the rotating frame of such a binary, the isopotential surfaces close to each of the stars will be approximately spherical, but at larger radii they are more and more oval shaped, due to the gravitational pull of the companion (see Fig. 4.12). There is one particular isopotential surface for which projections onto any plane passing through the line connecting the stars traces a "figure 8", i.e., the surface is pinched into two pointed "lobes" that connect at a point between the two stars. These are called **Roche lobes** and the point where they connect is the **first Lagrange point**, L_1. At L_1, the gravitational forces due to the two stars, and the centrifugal force in the rotating frame due to rotation about the center of mass, all sum up to zero.[10]

In any star, surfaces of constant gas density and pressure will be parallel to surfaces of constant potential (which is why isolated stars are spherical). Thus, a member of a close binary that evolves and grows in radius, e.g., into a red giant, will have a shape that

---

[9] The same kind of tidal deformation is applied by the Sun and the Moon to the Earth, especially to the Earth's liquid water surface layer. The deformation is maximal when the three bodies are approximately aligned, during full Moon and new Moon. During one daily Earth rotation, a point on the Earth goes through two "high tide" locations and two "low tide" locations. Due to the loss of energy to tidal friction, the Earth–Moon system is by now largely circularized, but only partly synchronized. On the one hand, the Moon's orbital and rotation periods are exactly equal, and hence we always see the same ("near") side of the Moon. Although the Moon is solid, synchronization was achieved by means of the solid tidal stresses and deformations imposed on it by the Earth. The Earth's rotation, on the other hand, is not yet synchronized with either the Sun's or the Moon's orbital periods. See Problem 8 for some quantitative assessments of ocean tides.

[10] Note that L_1 is generally *not* at the center of mass. The center of mass is closer to the more massive star in the binary system, while L_1 is closer to the less massive star. Only in equal-mass binaries do the two points coincide.

is increasingly teardrop shaped. If the star inflates enough to fill its Roche lobe, stellar material at the L_1 point is no longer bound to the star, and can fall onto the companion. Three configurations are thus possible:

In a **detached binary** neither of the stars fills its Roche lobe; in a **semi-detached binary** one of the stars fills its Roche lobe; and in a **contact binary** both stars fill their Roche lobes. In the last case the binary system looks like a single, peanut-shaped object with two stellar cores and a common envelope.

In the semi-detached case there is always transfer of matter from the Roche-lobe-filling star to its companion. Different observational phenomena result, depending on the nature of the receiving star. If it is a main-sequence star, an *Algol-type* binary system results. If the receiving star is a white dwarf, the resulting phenomena are called *cataclysmic variables*, *novae*, and *type Ia supernovae*. If the receiver is a neutron star or black hole, the system is called an *X-ray binary*.

Viewed from an inertial reference frame, the accreted material possesses angular momentum having the direction of the system's orbital angular momentum. (In the rotating frame, the matter experiences a Coriolis force as it falls toward the receiving star.) If the receiving star is compact, the accreted material will not reach the surface immediately, but rather go into orbit around the star. The gas particles on different coplanar, elliptical orbits will collide with each other, and eventually an **accretion disk** forms around the receiving star.

4.6.1 Accretion Disks

In an accretion disk, particles move on approximately circular orbits, and lose energy and angular momentum due to viscous interaction with particles moving along orbits at adjacent radii. The particles therefore slowly drift to progressively smaller radii, until reaching the surface of the star (or the Schwarzschild radius, if the accretor is a black hole). The frictional heat is radiated away. Although the exact process by which viscous friction operates in accretion disks is still a matter of debate, we can nonetheless derive some general properties of these objects.

Consider a mass element, dM, in the accretion disk around a star of mass M. To fall from a circular Keplerian orbit of radius $r + dr$ to an orbit at radius r, the mass element must lose some potential energy. Half of the lost potential energy is necessarily converted to additional kinetic energy at the smaller radius with its higher Keplerian velocity, and the remaining half can be converted to heat.[11] The gain in thermal energy of the mass element will thus be

$$dE_{\text{th}} = \frac{1}{2} \left(\frac{GMdM}{r} - \frac{GMdM}{r + dr} \right),$$

(4.128)

---

[11] Note that this result, while following directly from Newtonian mechanics for a particle in a circular orbit, is just another instance of the virial theorem for a classical nonrelativistic system of particles in gravitational equilibrium—in this case a system of one particle.

where we neglect the gravitational self-binding energy of the disk itself. Assuming that the hot gas radiates its thermal energy as a blackbody at the same radius where the gravitational energy is liberated, the luminosity from an annulus in the disk will be

$$dL = \frac{dE}{dt} = \frac{1}{2}GM\frac{dM}{dt}\left(\frac{1}{r} - \frac{1}{r+dr}\right) = \frac{1}{2}GM\dot{M}\frac{dr}{r^2} = 2(2\pi r)dr\sigma T^4, \tag{4.129}$$

where $\dot{M}$ is the mass accretion rate through a particular annulus of the disk, σ is the Stefan-Boltzmann constant, and the factor of 2 on the right-hand side is because the area of the annulus includes both the "top" and the "bottom." Taking the two right-hand terms and isolating T, we find for the temperature profile of an accretion disk

$$T(r) = \left(\frac{GM\dot{M}}{8\pi\sigma}\right)^{1/4}r^{-3/4}. \tag{4.130}$$

In a steady state, $\dot{M}$ must be independent of r (otherwise material would pile up in the disk, or there would be a shortage of material at small radii), and must equal the accretion rate of mass reaching the stellar surface. Thus, $T \propto r^{-3/4}$, meaning that the inner regions of the disk are the hottest ones, and it is from them that most of the luminosity emerges. The total luminosity of an accretion disk with inner and outer radii $r_{\rm in}$ and $r_{\rm out}$ is found by integrating over the luminosity from all annuli,

$$L = \int_{r_{\rm in}}^{r_{\rm out}} 2(2\pi r)\sigma T^4(r)dr = \frac{1}{2}GM\dot{M}\left(\frac{1}{r_{\rm in}} - \frac{1}{r_{\rm out}}\right). \tag{4.131}$$

This result could have, of course, been obtained directly from conservation of energy.[12] If $r_{\rm out} \gg r_{\rm in}$, the result simplifies further to

$$L = \frac{1}{2}\frac{GM\dot{M}}{r_{\rm in}}. \tag{4.132}$$

It is instructive to evaluate the **radiative efficiency** of accretion disks by dividing the luminosity above by $\dot{M}c^2$, the hypothetical power that would be obtained if all the accreted rest mass were converted to energy:

$$\eta = \frac{1}{2}\frac{GM}{c^2 r_{\rm in}}. \tag{4.133}$$

If the accreting object is, e.g., a $1.4M_\odot$ neutron star with an accretion disk reaching down to the stellar surface at a radius of 10 km, then $r_{\rm in}$ is about 2.5 times the Schwarzschild radius, $r_s = 2GM/c^2$ (Eq. 4.107), that corresponds to such a mass (recall that $r_s \approx 3$ km for

---

[12] Note that, in addition to energy conservation, a full treatment of accretion disk structure must also conserve angular momentum. The angular momentum per unit mass of a disk particle at radius r, in a circular Keplerian orbit with velocity v_c, is $J/m = rv_c = \sqrt{GMr}$. Thus, a particle descending to an orbit at smaller r must get rid of angular momentum by transfering it outward to other particles in the disk via viscous torques. Some particles at the outer edge of the disk must therefore gain angular momentum, and hence move to larger radii. Some of the gravitational energy released by the inflow will power this outflow of matter, at the expense of the energy that can be radiated by the disk. The work done by the frictional torques also increases by a factor of ≈3 the thermal energies in the outer radii of the disk, at the expense of the inner radii, slightly modifying Eq. 4.130. The exact form will depend also on the amount of angular momentum transferred to the accreting object at the inner edge of the disk.

$1M_\odot$). The rest-mass-to-radiative energy conversion efficiency is then about 0.10. For black-hole accretors, it turns out from solution of the general relativity equations of motion that gas particles have a *last stable orbit* at which they can populate the accretion disk. At smaller radii, a particle quickly spirals in and crosses the event horizon, carrying its remaining kinetic energy with it. The last stable orbit for a nonrotating black hole[13] is at $3r_s$. Accretion disks around such black holes will therefore have an efficiency of $1/12 \approx 0.08$, somewhat lower than accretion disks around neutron stars. (A solution of the problem using the correct general relativistic, rather than Newtonian, potential, gives an efficiency of 0.057). The point to note, however, is that, in either case, the efficiency is an order of magnitude higher than the efficiency of the nuclear reactions operating in stars, $\eta = 0.007$ or less. Furthermore, only a tiny fraction of a main-sequence star's mass is involved at any given time in nuclear reactions, whereas an accretion disk can extract energy with high efficiency from all of the mass being channeled through it. Under appropriate conditions, accretion disks can therefore produce high luminosities.

Let us calculate the typical luminosities and temperatures of accretion disks in various situations. In **cataclysmic variables**, the accretor is a white dwarf, with a typical mass of $1M_\odot$ and a radius of 10^4 km. A typical accretion rate[14] is $10^{-9}M_\odot$ yr^{-1}. This produces a luminosity of

$$L = \frac{1}{2}\frac{GM\dot{M}}{r_{\text{in}}} = \frac{6.7 \times 10^{-8} \text{ cgs} \times 2 \times 10^{33} \text{ g} \times 10^{-9} \times 2 \times 10^{33} \text{ g}}{2 \times 3.15 \times 10^7 \text{ s} \times 10^9 \text{ cm}}$$
$$= 4 \times 10^{33} \text{ erg s}^{-1} \approx L_\odot. \tag{4.134}$$

The luminosity from the accretion disk thus completely overpowers the luminosity of the white dwarf. The disk luminosity can be much greater than that of the donor star (for low-mass main-sequence donors, the most common case), comparable to the donor star (for intermediate-mass main sequence stars) or much smaller than the donor luminosity (for high-mass main sequence and red-giant donors). At the inner radius (which dominates the luminosity from the disk) the temperature is (Eq. 4.130)

$$T(r) = \left(\frac{GM\dot{M}}{8\pi\sigma}\right)^{1/4} r^{-3/4}$$
$$= \left(\frac{6.7 \times 10^{-8} \text{ cgs} \times 2 \times 10^{33} \text{ g} \times 10^{-9} \times 2 \times 10^{33} \text{ g}}{3.15 \times 10^7 \text{ s} \times 8\pi \times 5.7 \times 10^{-5} \text{ cgs}}\right)^{1/4} (10^9 \text{ cm})^{-3/4}$$
$$= 5 \times 10^4 \text{ K}. \tag{4.135}$$

---

[13] A black hole is fully characterized by only three parameters—its mass, its spin angular momentum, and its electric charge (the latter probably not being of astrophysical relevance, because astronomical bodies are expected to be almost completely neutral). Spacetime around a rotating black hole is described by a metric called the Kerr metric, rather than by the Schwarzschild metric. Black-hole spin is accompanied by the general relativistic phenomenon of "frame dragging," in which spacetime outside the event horizon rotates with the black hole. In a rotating black hole, the last stable orbit and the event horizon are at smaller radii than in the nonrotating case.

[14] The accretion rate can be limited by the rate at which the donor star transfers mass through the L_1 point, by the efficiency of the viscous process that causes material in the accretion disk to fall to smaller radii, or by the radiation pressure of the luminosity resulting from the accretion process—see section 4.6.2.

The thermal spectrum from the disk therefore peaks in the far UV part of the spectrum, and is usually different from the spectrum of the main-sequence or red-giant donor star (which of course generally has a red spectrum). The integrated spectrum of the system will therefore have at least two distinct components.

When the orbits of cataclysmic variables are sufficiently inclined to our line of sight, monitoring the total light output over time, as the systems rotate, reveals changes due to mutual eclipses by the various components: the donor star, the accretion disk, and sometimes a "hot spot" where the stream of matter from the donor hits the disk. The changing projected area of the distorted donor star also affects the light output. Analysis of such data allows reconstructing the configurations and parameters of these systems. In addition to the periodic variability induced by eclipses and changes in orientation, accreting systems reveal also aperiodic variability, i.e., variations with a "noise-like" character. These variations likely arise from an unstable flow of the material overflowing the donor's Roche lobe, causing changes in $\dot{M}$, as well as from instabilities and flares in the accretion disk itself.

In a class of cataclysmic variable called **novae** there are also outbursts of luminosity during which the system brightens dramatically for about a month. The outbursts occur once every 10–10^5 yr, as a result of rapid thermonuclear burning of the hydrogen-rich (and hence potentially explosive) accreted material that has accumulated on the surface of the white dwarf. Assuming again an accretion rate of $10^{-9} M_\odot \, \mathrm{yr}^{-1}$, over a period of 1000 yr, a mass of $10^{-6} M_\odot$ will cover the surface of the white dwarf. If completely ignited, it yields an energy

$$E_{\mathrm{nova}} = 0.007 mc^2 = 0.007 \times 10^{-6} \times 2 \times 10^{33} \, \mathrm{g} \times (3 \times 10^{10} \, \mathrm{cm \, s^{-1}})^2 \approx 10^{46} \, \mathrm{erg}. \quad (4.136)$$

When divided by a month (2.5×10^6 s), this gives a mean luminosity of $4 \times 10^{39} \, \mathrm{erg \, s^{-1}} = 10^6 L_\odot$, i.e., 10^6 times the normal luminosity of the accretion disk. In reality, only partial processing of the accreted hydrogen takes place, and the energy is also partly consumed in unbinding some material from the underlying white dwarf. On the other hand, for longer recurrence times between outbursts, the mass of accumulated hydrogen can be larger than assumed above. The gamma-ray spectra of novae reveal emission from the radioactive decay of elements that are synthesized in these explosions, providing direct evidence of the process at hand.

As discussed in section 4.3.3, under certain conditions (likely involving the reaching of the Chandrasekhar mass by the accreting white dwarf) an extreme, runaway version of the nova eruption, called a **type Ia supernova**, occurs. In such an event, a large fraction of the white dwarf mass (i.e., of order $1 M_\odot$ of carbon, rather than the $10^{-6} M_\odot$ of hydrogen in the nova case) is ignited and is explosively synthesized into iron-group elements. The total energy is, correspondingly, 10^6 times larger than that of a nova, i.e., 10^{51-52} erg. As in the core-collapse supernova explosions that end the life of massive stars, the ratio of kinetic to luminous energy is about 100, and thus type Ia supernovae, with a luminosity of about $10^{10} L_\odot$, can outshine their host galaxies for a period of about a month (see Problem 4). Although core-collape supernovae and type Ia supernovae have similar luminous and kinetic energy outputs, one should remember that in core-collapse supernovae

99% of the energy is carried away by neutrinos, and therefore core-collapse supernovae are intrinsically far more energetic events. Type Ia supernovae have a narrow range of observed optical luminosities, probably as a result of the fact that they generally involve the explosion of about $1.4 M_\odot$ of white dwarf material. These supernovae are therefore very useful as "standard candles" for measuring distances. In chapters 7 and 9 we will see how they have been used in this application.

When the receiving star in an interacting binary is a neutron star or a black hole, the inner radius of the accretion disk is of order 10 km, rather than 10^4 km, and therefore the luminosity is much greater than in a white-dwarf accretor. For example, scaling from Eq. 4.134, if the accretor is a $1.4 M_\odot$ neutron star with the same accretion rate, the accretion-disk luminosity is of order 10^{37} erg s^{-1}. The temperature at the inner radius, scaling as $M^{1/4} r^{-3/4}$ (Eq. 4.135), is $T = 10^7$ K. The emission therefore peaks in the X-rays, and hence the name **X-ray binaries**. In reality, due to the extreme matter and radiation densities, temperatures, and magnetic fields near the surface of a neutron star, the accretion disk may not actually reach the surface, and accreting material is sometimes channeled to the poles, forming a hot-spot where it hits the surface. In addition to the thermal emission from the accretion disk, other, nonthermal, radiation components are observed in such systems, e.g., synchrotron emission from relativistic electrons spiraling along magnetic field lines. Some accreting white dwarfs also possess strong magnetic fields that funnel the accretion flow directly onto hot spots on the white dwarf. Such *magnetic cataclysmic variables* also appear then as X-ray sources.

4.6.2 Accretion Rate and Eddington Luminosity

The above discussion shows that the properties of accreting systems are largely determined by three parameters, M, $\dot{M}$, and $r_{\rm in}$. M and $r_{\rm in}$ are limited to particular values by the properties of stars and stellar remnants. However, the accretion rate, $\dot{M}$, also cannot assume arbitrarily large values. To see this, consider an electron at a radius r in an ionized gas that is taking part in an accretion flow toward some compact object of mass M. The accretion flow produces a luminosity per frequency interval L_ν, and therefore the *density of photons* with energy $h\nu$ at r is

$$n_{\rm ph} = \frac{L_\nu}{4\pi r^2 c h\nu}. \tag{4.137}$$

The rate at which photons of this energy are scattered via Thomson scattering on the electron is

$$R_{\rm scat} = n_{\rm ph} \sigma_T c, \tag{4.138}$$

where σ_T is the Thomson scattering cross section. Each scattering event transfers, on average, a momentum $p = h\nu/c$ to the electron. The rate of momentum transfer to the electron, i.e., the force exerted on it by the radiation, is then

$$\frac{dp}{dt} = R_{\rm scat} \frac{h\nu}{c} = \frac{L_\nu \sigma_T}{4\pi r^2 c}. \tag{4.139}$$

The total radiative force on the electron is obtained by integrating over all frequencies ν,

$$F_{\text{rad}} = \frac{L\sigma_T}{4\pi r^2 c}.$$ (4.140)

The electron would be repelled from the accreting source of luminosity, were it not for the gravitational attraction of the accreting object. This force will be much greater on protons than on electrons. However, the Coulomb attraction between electrons and protons prevents their separation, and therefore the gravitational attraction on a proton effectively operates on neighboring electrons as well. The attractive force on the electron is therefore

$$F_{\text{grav}} = \frac{GMm_p}{r^2}.$$ (4.141)

The accretion flow, and its resulting luminosity, can proceed only if the radiative force does not halt the inward flow of matter, i.e., $F_{\text{rad}} < F_{\text{grav}}$. Equating the two forces, using Eqs. 4.140 and 4.141, we obtain the maximum luminosity possible in a system powered by accretion,

$$L_E = \frac{4\pi cGMm_p}{\sigma_T}$$ (4.142)

$$= \frac{4\pi \times 3 \times 10^{10} \times 6.7 \times 10^{-8} \text{ cgs} \times 2 \times 10^{33} \text{ g} \times 1.7 \times 10^{-24} \text{ g}}{6.7 \times 10^{-25} \text{ cm}^2} \frac{M}{M_\odot}$$

$$= 1.3 \times 10^{38} \text{ erg s}^{-1} \frac{M}{M_\odot} = 6.5 \times 10^4 \, L_\odot \frac{M}{M_\odot}.$$

This limiting luminosity is called the **Eddington luminosity**.

Recalling our derivation, above, of a luminosity of order 10^{37} erg s^{-1} from an accretion disk around a $1.4M_\odot$ neutron star with an accretion rate $\dot{M} = 10^{-9}M_\odot$ yr^{-1}, we see that an accretion rate, say, 100 times larger would bring the system to a luminosity of several times L_E, and is therefore impossible. This is not strictly true, since in the derivation of L_E we have assumed spherical accretion and an isotropically radiating source. Both assumptions fail in an accretion disk, which takes in matter along an equatorial plane, and radiates preferentially in directions perpendicular to that plane. Nevertheless, detailed models of accretion disk structure show that disks become unstable when radiating at luminosities approaching L_E. The Eddington limit is therefore a useful benchmark even for nonspherical accreting systems. Finally, note that L_E applies to systems undergoing steady-state accretion. Objects of a given mass can have higher luminosities (see, e.g., the luminosities of novae and supernovae that we calculated above), but then an outflow of material is unavoidable, the object is disrupted, and the large luminosity must be transient.

4.6.3 Evolution of Interacting Binary Systems

The transfer of mass between members of interacting binaries can have drastic effects on both members. We recall that isolated neutron stars power their pulsar emission and their surrounding supernova remnant emission at the expense of their rotational energy, and thus gradually slow down. A neutron star in a binary system, if accreting matter from its companion, under suitable conditions can gain angular momentum, which can spin the pulsar back up. Thus, many pulsars in binary systems are spinning at millisecond frequencies, i.e., close to the maximal spin possible for a neutron star, and have negative period derivatives, $\dot{P}$ (if they are still being spun up by the accretion; see Problem 9). The neutron star can also affect the donor star. The jets and beams present in pulsars may hit one side of the donor star (the binaries are tidally locked), heat it, ablate it, or completely destroy it. Several examples of such *black-widow pulsars* are known, in which an old millisecond pulsar has no companion, or in which the companion is a white dwarf of much too small a mass to have evolved in isolation from the main sequence (i.e., white dwarfs of such mass form after a time that is much greater than the age of the Universe).

The transfer of mass and angular momentum in an interacting binary can also lead to complex evolution of the parameters of the system, such as binary separation and accretion rate. Changes in those parameters can then affect the future evolution of the system. Let us see how this works. The orbital angular momentum of a circular binary composed of masses M_1 and M_2 with separation a is

$$J = I\omega = \mu a^2 \omega, \tag{4.143}$$

where I is the moment of inertia, and μ is the reduced mass,

$$\mu = \frac{M_1 M_2}{M_1 + M_2}. \tag{4.144}$$

(For simplicity, we will ignore the spin angular momentum of the stars.) Substituting ω from Kepler's law (Eq. 2.35),

$$\omega^2 = \frac{G(M_1 + M_2)}{a^3}, \tag{4.145}$$

we get

$$J = \mu\sqrt{G(M_1 + M_2)a}. \tag{4.146}$$

Assuming conservation of total mass and angular momentum, the time derivative of J equals zero,

$$\frac{dJ}{dt} = \sqrt{G(M_1 + M_2)}\left(\frac{d\mu}{dt}\sqrt{a} + \frac{\mu}{2\sqrt{a}}\frac{da}{dt}\right) = 0, \tag{4.147}$$

or

$$-\frac{2}{\mu}\frac{d\mu}{dt} = \frac{1}{a}\frac{da}{dt}. \tag{4.148}$$

Expressing $\dot{\mu}$ in terms of its constituent masses gives

$$\frac{d\mu}{dt} = \frac{1}{M_1 + M_2} \left(\frac{dM_1}{dt} M_2 + M_1 \frac{dM_2}{dt} \right). \tag{4.149}$$

However, conservation of mass means that $\dot{M}_1 = -\dot{M}_2$, and hence

$$\frac{d\mu}{dt} = \frac{\dot{M}_1}{M_1 + M_2} (M_2 - M_1). \tag{4.150}$$

Replacing in Eq. 4.148, we finally get

$$2\dot{M}_1 \frac{M_1 - M_2}{M_1 M_2} = \frac{1}{a} \frac{da}{dt}. \tag{4.151}$$

Equation 4.151 determines how the period and separation of the system evolve, depending on the constituent masses, the accretion rate, and its sign. For example, consider a system that starts out with two close main sequence stars, with $M_1 > M_2$. M_1 will therefore be the first to become a red giant, fill its Roche lobe, and transfer mass to M_2. Since M_1 loses mass, $\dot{M}_1$ is negative. From Eq. 4.151, $\dot{a}$ is then negative. In other words, the two stars approach each other. The decrease in separation a means that the Roche lobe around M_1 moves to a smaller radius, and the accretion rate grows further. If this trend is not interrupted (e.g., by the end of the giant stage of M_1), the system reaches a common envelope stage. Evolution resumes once M_1 becomes a white dwarf, or at a later stage, when M_2 becomes a red giant, if it fills its Roche lobe. Accretion will now be in the opposite sense, and $\dot{M}_1$ is therefore positive. If, despite the earlier accretion phase and the individual stellar evolution, M_1 is still larger than M_2, then $\dot{a}$ will now be positive. If the Roche lobe size of M_2 overtakes the star's radius, accretion will stop. Alternatively, if by this time $M_2 > M_1$, the two stars will again approach each other and there may be a second common envelope phase. Obviously, there are many other possible evolution paths, depending on the initial parameters. Moreover, in reality stars lose mass throughout their evolution by means of winds, and therefore the total mass and angular momentum of a binary system will generally not be conserved, opening further binary evolution paths.

Problems

1. In a fully degenerate gas, all the particles have energies lower than the Fermi energy. For such a gas we found (Eq. 4.19) the relation between the density n_e and the Fermi momentum p_f:

$$n_e = \frac{8\pi}{3h^3} p_f^3.$$

   a. For a nonrelativistic electron gas, use the relation $p_f = \sqrt{2m_e E_f}$ between the Fermi momentum, the electrom mass m_e, and the Fermi energy E_f, to express E_f in terms of n_e and m_e.

b. Estimate a characteristic n_e under typical conditions inside a white dwarf. Using the result of (a), and assuming a temperature $T = 10^7$ K, evaluate numerically the ratio E_{th}/E_f, where E_{th} is the characteristic thermal energy of an electron in a gas of temperature T, to see that the electrons inside a white dwarf are indeed degenerate.

2. Cold, planetary-mass objects such as Jupiter are mostly devoid of internal thermal energy sources, as is the case of white dwarfs. However, planets are supported against gravity by repulsive atomic electrostatic forces rather than by free electron degeneracy pressure. Estimate the maximum mass that can be supported by atomic electrostatic forces, as follows.

a. Approximate the typical pressure inside a planet by means of the electrostatic Coulomb energy density due to each atom's repulsion of its adjacent atoms. Ignore the effect of nonadjacent atoms, whose charges are screened. Assume a pure atomic-hydrogen composition. Assume further that the atoms are distributed on a static grid of constant density with separations r, and hence there are six neighboring atoms surrounding each atom, with the centers of their electron clouds separated by $\sim r$ from the center of the electron cloud of the central atom.

b. Express the planet radius in terms of the planet mass M, the hydrogen atom mass, m_H, and the "rigid" interatomic spacing r, and then write the gravitational binding energy density of the planet in terms of these parameters.

c. Equate the electrostatic energy density you found to the gravitational binding energy density. The interatomic spacing r should cancel out from the equation (why?). Find the mass at which this equality occurs, and compare to Jupiter's mass, $M_J \approx 0.001 M_\odot$.

Answer: $8M_J$. For larger masses, the gravitational energy density will overcome the atomic electrostatic repulsion, the planet radius will stop growing with mass as fast as $M^{1/3}$, the density will increase, and quantum degeneracy pressure of the electrons will set in as the main source of pressure. From there on, the planet's radius will decrease as its mass increases, as $M^{-1/3}$ (Eq. 4.34).

3. Most of the energy released in the collapse of a massive star to a neutron star (a core-collapse supernova) is in the form of neutrinos.

a. If the just-formed neutron star has a mass $M = 1.4M_\odot$ and a radius $R = 10$ km, estimate the mean nucleon density, in cm^{-3}. Find the mean free path, in cm, of a neutrino inside the neutron star, assuming the density you found and a cross section for scattering of neutrinos on neutrons of $\sigma_{vn} = 10^{-42}$ cm^2.

b. How many seconds does it take a typical neutrino to emerge from the neutron star in a random walk?

Hint: Neutrinos travel at a velocity close to c. Recall that the radial distance d covered in a random walk of N steps, each of length l, is $d = \sqrt{N}l$.

c. Twelve electron antineutrinos from Supernova 1987A were detected by the Kamiokande neutrino detector in Japan. This experiment consisted of a tank filled with 3 kton of water, and surrounded by photomultiplier tubes. The photomultipliers detect the Cerenkov radiation emitted by a recoiling positron that is emitted after a proton absorbs an antineutrino from the supernova. Estimate how many people on Earth could have perceived a flash of light, due to the Cerenkov radiation produced by the same process, when an antineutrino from the supernova traveled through their eyeball. Assume that eyeballs are composed primarily of water, each weighs about 10 g, and that the Earth's population was 5 billion in 1987.

4. Type Ia supernovae are probably thermonuclear explosions of accreting white dwarfs that have approached or reached the Chandrasekhar limit.
   a. Use the virial theorem to obtain an expression for the mean pressure inside a white dwarf of mass M and radius R.
   b. Use the result of (a) to estimate, to an order of magnitude, the speed of sound, $v_s = \sqrt{dP/d\rho} \sim \sqrt{P/\rho}$, inside a white dwarf. In an accreting white dwarf with a carbon core that has reached nuclear ignition temperature, a nuclear burning "flame" encompasses the star at the sound velocity or faster. Within how much time, in seconds, does the flame traverse the radius of the white dwarf, assuming $R = 10^4$ km, $M = 1.4 M_\odot$? Note that this **sound-crossing timescale** is $\sim (G\rho)^{-1/2}$, which is also the free-fall timescale (Eq. 3.15.)
   c. Calculate the total energy output, in ergs, of the explosion, assuming that the entire mass of the white dwarf is synthesized from carbon to nickel, with a mass-to-energy conversion efficiency of 0.1%. Compare this energy to the gravitational binding energy of the white dwarf, to demonstrate that the white dwarf explodes completely, without leaving any remnant.
   d. Gamma rays from the radioactive decays $^{56}Ni \rightarrow ^{56}Co + \gamma \rightarrow ^{56}Fe + \gamma$ drive most of the optical luminosity of the supernova. The atomic weights of ^{56}Ni and ^{56}Fe are 55.942135 and 55.934941, respectively. Calculate the total energy radiated in the optical range during the event. Given that the characteristic times for the two radioactive decay processes are 8.8 days and 111 days, respectively, show that the typical luminosity is $\sim 10^{10} L_\odot$.

5. General relativity predicts that accelerated masses radiate **gravitational waves**, thereby losing energy, in analogy to the emission of electromagnetic radiation by accelerated charges. There is indirect evidence for the existence of such waves from the orbital time evolution of some binary pulsars. If gravitational radiation were also responsible for the loss of rotational energy E_{rot} of isolated pulsars (e.g., the Crab pulsar), then a dependence

$$\frac{dE_{rot}}{dt} \propto \omega^6$$

would be expected, where ω is the angular rotation velocity.
   a. Under the above assumption, find an expression for $\omega(t)$.
   b. For the time dependence found in (a), derive an upper limit for the age of the Crab pulsar. Given that the supernova that marked the Crab's formation occurred in the year 1054, is gravitational radiation a viable braking mechanism for the Crab pulsar?

6. A type Ia supernova is thought to be the thermonuclear explosion of an accreting white dwarf that goes over the Chandrasekhar limit (see Problem 4). An alternative scenario, however, is that supernova Ia progenitors are white dwarf binaries that lose orbital energy to gravitational waves (see Problem 5) until they merge, and thus exceed the Chandrasekhar mass and explode.

   a. Show that the orbital kinetic energy of an equal-mass binary with separation a and individual masses M is

$$E_k = \frac{GM^2}{2a},$$

   and the total orbital energy (kinetic plus gravitational) is minus this amount.

   b. The power lost to gravitational radiation by such a system is

$$\dot{E}_{gw} = -\frac{2c^5}{5G}\left(\frac{2GM}{c^2 a}\right)^5.$$

   By equating to the time derivative of the total energy found in (a), obtain a differential equation for $a(t)$, and solve it.

   c. What is the maximum initial separation that a white-dwarf binary can have, if the components are to merge within 10 Gyr? Assume the white dwarfs have $1 M_\odot$ each, and the merger occurs when $a = 0$.
   Answer: 0.016 AU.

7. A star of mass m and radius r approaches a black hole of mass M to within a distance $d \gg r$.

   a. Using Eq. 4.127, express, in terms of m, r, and M, the distance d at which the Newtonian radial tidal force exerted by the black hole on the star equals the gravitational binding force of the star, and hence the star will be torn apart.

   b. Find the black-hole mass M above which the tidal disruption distance, d, is smaller than the Schwarzschild radius of the black hole, and evaluate it for a star with $m = M_\odot$ and $r = r_\odot$. Black holes with masses above this value can swallow Sun-like stars whole, without first tidally shredding them.
   Answer: $1.6 \times 10^8 M_\odot$.

   c. Derive a Newtonian expression for the **tangential** tidal force exerted inward on the star, in terms of m, r, M, and d, again under the approximation $r \ll d$. The combined effects of the radial tidal force in (a) and and the tangential tidal force in (c) will lead to "spaghettification" of stars, or other objects that approach the black hole to within the disruption distance.

   Hint: Remember that the star is in a *radial* gravitational field, and hence there is a tangential component to the gravitational force exerted on regions of the star that are off the axis defined by the black hole and the center of the star. The tangential component can be found by noting that the small angle between the axis and the edge of the star is $\approx r/d$.

8. Practitioners of some schools of yoga are warned not to perform yoga during the full or the new Moon, citing the tidal effect of the Moon at those times on other "watery bodies" such as the oceans. Let us investigate this idea.

   a. Verify the dramatic tidal effect of the Moon on the oceans by using Eq. 4.127 to calculate the ratio of the tidal "lifting force" F_{tide} and the Earth's gravitational attraction F_{grav}, for a point mass on the surface of the Earth and on the Earth–Moon line. Use this ratio to estimate the change in water height, in cm, between high tide and low tide, due to the moon alone. Repeat the calculation for the tidal effect due to the Sun alone.

   Hint: The surface of the oceans traces an equipotential surface, $gR = $ constant, where R is the distance from the Earth's center to the ocean surface at every point, and g is the effective gravitational acceleration at every point on the surface. Translate the tidal-to-gravitational force ratio into a relative change in g between a point at high tide (which experiences the full tidal force) and a point at low tide (which experiences no tidal force), and thus derive the relative change in R.

   Answers: 77 cm due to Moon, 33 cm due to Sun.

   Note: While the Sun and Moon are the drivers of ocean tides, a reliable calculation of tides at a particular Earth location must take into account additional factors, including the varying distances between Earth, Moon, and Sun (due to their elliptical orbits), their inclined orbital planes, the latitude, coastline shape, beach profile, ocean depth, water viscosity and salinity, and prevailing ocean currents.

   b. Calculate by how much (in milligrams) you are lighter when the full or new Moon is overhead or below, compared to when it is rising or setting, assuming your body weight is 50 kg.

   c. Calculate in dynes and in gram-force (i.e., in dynes divided by the gravitational acceleration $g = 980$ cm s^{-2}) the tidal **stretch** exerted on your body by the Moon plus the Sun when you are standing up with the full or new moon overhead. Assume your body weight is 50 kg and your height is 180 cm.

9. A spinning neutron star of mass $M = 1.4\, M_\odot$, constant density, and radius $R = 10$ km has a period $P = 1$ s. The neutron star is accreting mass from a binary companion through an accretion disk, at a rate of $\dot{M} = 10^{-9}\, M_\odot$ yr^{-1}. Assume that the accreted matter is in a circular Keplerian orbit around the neutron star until just before it hits the surface, and once it does then all of the matter's angular momentum is transferred onto the neutron star.

   a. Derive a differential equation for $\dot{P}$, the rate at which the neutron-star period decreases.

   b. Solve the equation to find how long will it take to reach $P = 1$ ms, the maximal spin rate of a neutron star.

   Hint: Calculate the Keplerian velocity of the accreted material a moment before it hits the neutron star surface, and use it to derive the angular momentum per unit mass of this material, J/m. The angular momentum of a rotating object with moment of inertia I is $I\omega$. The rate of change of the star's angular momentum is just the rate at which it receives angular momentum from the accreted matter, i.e.,

$$\frac{d}{dt}(I\omega) = \dot{M}\frac{J}{m}.$$

The moment of inertia of a constant-density sphere is $I = \frac{2}{5}MR^2$. Solve for the angular acceleration $\dot{\omega}$, neglecting changes in the neutron star's mass and radius. (This will be justified by the result.) From the relation $P = 2\pi/\omega$, derive $\dot{P}$. This "spin-up" process explains the properties of old, "millisecond pulsars," some of which, indeed, have negative $\dot{P}$.

Answer: 2.6×10^8 yr. Over this time, the neutron star mass increases by 18%, and its radius decreases by 5%, justifying the approximation of constant mass and radius.

10. A compact accreting object of mass M is radiating at the Eddington luminosity (Eq. 4.142) corresponding to that mass,

$$L_E = \frac{4\pi cGMm_p}{\sigma_T} = 1.3 \times 10^{38} \text{ erg s}^{-1}\frac{M}{M_\odot}.$$

An astronaut wearing a white space suit is placed at rest at an arbitrary distance from the compact object. Assuming that the cross-sectional area of the astronaut's body is $A = 1.5$ m^2, find the maximum allowed mass m of the astronaut, in kg, if the radiation pressure is to support her from falling onto the compact object.

Hint: By definition, a proton at any distance from this object will float, its gravitational attraction to the object balanced by the radiation pressure on nearby electrons. Consider the astronaut as a particle with mass m and cross section equal to her geometrical cross section, $2A$ (the factor of 2 is because her suit is white, so every photon reflection transfers twice the photon momentum). Compare m to m_p and $2A$ to σ_T.

Answer: 77 kg.

5 | Star Formation, H ɪɪ Regions, and the Interstellar Medium

Much of the **baryonic** matter (i.e., matter composed of protons and neutrons) in the Universe is not inside stars, but is, instead, distributed between the stars. The **interstellar medium** (ISM) consists of molecular, atomic, and ionized gas, with large ranges of densities and temperatures, as well as solid particles of **dust**. Most of the volume of the ISM is in the ionized and atomic gas phases. However, the molecular gas phase is particularly interesting because new generations of stars can form, under some conditions, out of this material. The newly formed stars then influence the remainder of the gas from which they formed, by means of outflows, photodissociation and photoionization by the radiation from massive stars, shocks and element enrichment from the explosions of the massive stars as core-collapse supernovae, and more. The ISM, in turn, releases this stellar energy by means of physical processes that are distinct from those we have seen in stars and stellar remnants. In this chapter, we focus on the ISM in star-forming regions. We briefly survey star forming environments, some of the properties of young stellar populations, the main physical processes that operate in the ISM around newly formed stars, and the observable consequences of those processes.

5.1 Cloud Collapse and Star Formation

Stars are seen to form in **molecular clouds**, which are one component of the ISM. Molecular clouds are randomly shaped agglomerations composed mainly of molecular-hydrogen gas, with typical masses of 10^2–$10^5 M_\odot$, sizes of hundreds of parsecs, and typical particle densities of 10^2–10^4 cm^{-3}. The temperatures in molecular clouds are between $\sim$10 and 100 K. Molecular clouds are the largest, most massive, gravitationally bound objects in the ISM. While the densities of molecular clouds are among the highest encountered in the ISM, it is important to realize that even this gas is extremely rarified compared to gas at an atmospheric density of $\sim$$10^{19}$ cm^{-3}. In fact, the densities of molecular clouds are many orders of magnitude lower than the density of the best vacua achievable in the laboratory.

(We will see below that these low densities have an important effect on the radiation emitted by the ISM.) The nearest (and probably the best-studied) giant molecular clouds are in the Orion star-forming region, at a distance of about 500 pc. Several views of this region are shown in Fig. 5.1, at various wavelengths and scales.

As we have seen, the mean mass densities inside stars are ~ 1 g cm^{-3}, i.e., particle densities of $\sim 10^{24}$ cm^{-3}. Thus, to form new stars, some regions of a molecular cloud must be compressed by many orders of magnitude. The details of the process of star formation are not understood yet. We can outline, however, some of the general criteria under which gravitational contraction of a gas cloud can proceed, and potentially lead to the conditions required for star formation.

5.1.1 The Jeans Instability

Let us assume, for simplicity, a spherical gas cloud of constant density ρ and temperature T, composed of particles with mean mass $\bar{m}$. The gas is ideal, classical, and nonrelativistic. The cloud's mass is M, its radius is r, and its gravitational energy is

$$|E_{\text{gr}}| \approx \frac{GM^2}{r}. \tag{5.1}$$

If the cloud undergoes a radial compression dr, the gravitational energy will change (become more negative) by

$$|dE_{\text{gr}}| = \frac{GM^2}{r^2} dr. \tag{5.2}$$

The volume will decrease by

$$dV = 4\pi r^2 dr, \tag{5.3}$$

and the thermal energy will therefore grow by

$$dE_{\text{th}} = PdV = nkT4\pi r^2 dr = \frac{M}{\bar{m}\frac{4}{3}\pi r^3}kT4\pi r^2 dr = 3\frac{M}{\bar{m}}kT\frac{dr}{r}. \tag{5.4}$$

The cloud will be unstable to gravitational collapse if the change in gravitational energy is greater than the rise in thermal energy (and the pressure support it provides),

$$|dE_{\text{gr}}| > dE_{\text{th}}. \tag{5.5}$$

Substituting from Eqs. 5.2 and 5.4, we see that this means that, for a cloud of given radius r and temperature T, collapse will occur if the mass is greater than the **Jeans mass**,

$$M_J = \frac{3kT}{G\bar{m}}r. \tag{5.6}$$

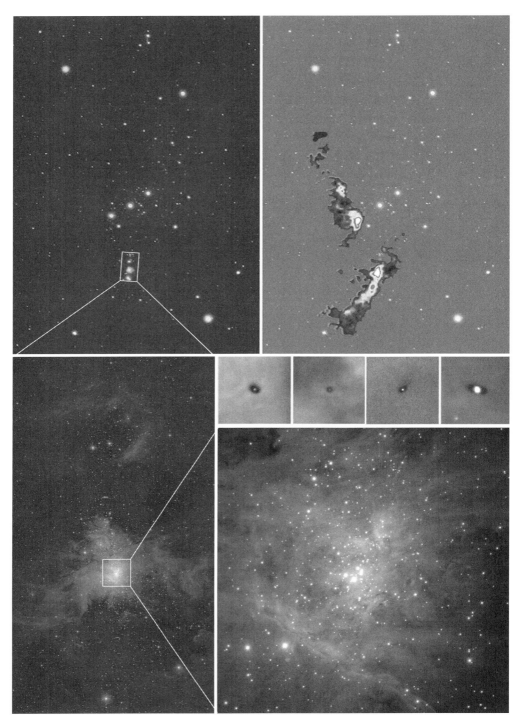

Figure 5.1 Several views of the Orion star-forming region. *Counterclockwise from top pair:* A map showing radio emission from the CO molecule, superimposed on an optical image of the familiar constellation; near-infrared zoom in on the region that constitutes the "sword" of the constellation, centered on the Orion nebula; a further zoom-in (about 1 pc on a side) on the "Trapezium" cluster of massive young stars that power the brightest H II region; and optical images of several protoplanetary disks around young stars, seen as dark silhouettes on the backdrop of the gas emission. Disk diameters are ∼100–500 AU. Photo credits: S. Sakamoto et al., see *Astrophys. J.*, 425 (1994), 641; G. Kopan, R. Hurt, and the Two Micron All Sky Survey; M. McCaughrean and the European Southern Observatory; and C. R. O'Dell—Vanderbilt University, ESA, and NASA.

Equivalently, a cloud of mass M will collapse only if its radius is smaller than the **Jeans radius**,

$$r_J = \frac{G\bar{m}}{3kT}M. \tag{5.7}$$

This condition can also be stated as a condition on the density, which must be larger than the **Jeans density**,

$$\rho_J = \frac{M}{\frac{4}{3}\pi r_J^3} = \frac{3}{4\pi M^2}\left(\frac{3kT}{G\bar{m}}\right)^3. \tag{5.8}$$

Let us obtain some representative numbers for a typical molecular cloud with $M \sim 1000 M_\odot$ and $T \sim 20$ K. At these low temperatures, there are few free electrons, and most of the hydrogen is bound in H_2 molecules, so $\bar{m} \approx 2m_H$. The Jeans density is then

$$\rho_J = \frac{3}{4\pi\,(1000 \times 2 \times 10^{33}\,\text{g})^2}\left(\frac{3 \times 1.4 \times 10^{-16}\,\text{erg K}^{-1} \times 20\,\text{K}}{6.7 \times 10^{-8}\,\text{cgs} \times 2 \times 1.7 \times 10^{-24}\,\text{g}}\right)^3$$

$$= 3 \times 10^{-24}\,\text{g cm}^{-3}, \tag{5.9}$$

corresponding to a number density of

$$n_J(H_2) = \frac{\rho_J}{\bar{m}} = \frac{3 \times 10^{-24}\,\text{g cm}^{-3}}{2 \times 1.7 \times 10^{-24}\,\text{g}} \sim 1\,\text{cm}^{-3}. \tag{5.10}$$

Thus, the typical observed density of molecular clouds, 10^2–10^4 cm^{-3}, is several orders of magnitude higher than the Jeans density and, according to the criterion we have just formulated, the clouds should be unstable to gravitational collapse. Since the clouds exist and appear to be long-lived, another source of pressure, other than thermal pressure, must be present. It is currently believed that the dominant pressure is provided by turbulence, magnetic fields, or both.

If, rather than looking at the entire $1000 M_\odot$ cloud, we consider a one-solar-mass clump of gas inside the cloud (with the same temperature of $T = 20$ K and density of 10^2–10^4 cm^{-3}), the Jeans density for this clump is 10^6 times larger, i.e., $\sim 10^6$ cm^{-3}. This is much higher than the actual density of the clump, and the clump is thus stable against collapse. We can therefore expect the collapse of a molecular cloud to begin with the larger masses, that more easily pass the Jeans criterion. As the collapse proceeds and the density increases, progressively smaller regions of the cloud reach the Jeans density for their masses, and can start to collapse independently. This fragmentation into smaller and denser masses continues until the creation of a cluster of individual **protostars**, i.e., objects that will evolve into stars on the main sequence. From Eq. 5.7, the Jeans radius of a molecular cloud fragment of mass of order $1M_\odot$, which can eventually collapse to such a protostar, is

$$r_J \sim 5 \times 10^{16}\,\text{cm}. \tag{5.11}$$

5.1.2 Cloud Collapse

Although the Jeans criterion provides a necessary condition for the onset of collapse in a gas cloud, the collapse involves the release of gravitational energy, and can proceed only if there is an avenue for releasing this energy. If, for example, the gravitational energy is converted to thermal energy and the temperature and the pressure rise, the Jeans mass will rise and the collapse will stop. The contraction will then proceed slowly, at a pace determined by the rate at which thermal energy is radiated away. However, if the gravitational energy is converted to some other, non-pressure-producing, form, then the collapse of a cloud that passes the Jeans criterion can proceed on a free-fall time. Two such avenues for converting gravitational energy, which are important in star-forming regions, are **dissociation** of H_2, which uses up 4.5 eV per molecule, and ionization of hydrogen, which takes 13.6 eV per atom.

Consider, for example, the $1M_\odot$ clump discussed above. If hydrogen dissociation and ionization can take place, then once the density has reached the Jeans value for such a mass, $\rho_J = 3 \times 10^{-18}$ g cm^{-3}, the cloud can collapse on a free-fall timescale (Eq. 3.15):

$$\tau_{ff} \sim \left(\frac{3\pi}{32G\rho}\right)^{1/2} = \left(\frac{3\pi}{32 \times 6.7 \times 10^{-8} \text{ cgs} \times 3 \times 10^{-18} \text{ g cm}^{-3}}\right)^{1/2}$$

$$= 1.2 \times 10^{12} \text{ s} = 40,000 \text{ yr}. \tag{5.12}$$

The free-fall collapse will stop once most of the hydrogen is first dissociated and then ionized, because continued compression past that point will quickly raise the temperature and pressure. We can find the properties of the collapsed cloud at its new equilibrium state from the virial theorem, which we developed in the context of stellar physics. At the new virial equilibrium, the thermal energy of the (now ionized) gas will again equal minus-one-half the gravitational energy. If we make the crude approximation that, during the collapse, the gas does not radiate away any energy (as in the treatment of convection in section 3.12, this is an *adiabatic* approximation), then minus the other half of the gravitational energy, in turn, equals the energy converted in breaking up the molecules and the atoms. Thus,

$$\frac{1}{2}E_{gr} = -\left(\frac{M}{2m_H}4.5 \text{ eV} + \frac{M}{m_H}13.6 \text{ eV}\right) \tag{5.13}$$

and

$$E_{th} = \frac{3}{2}NkT = \frac{3}{2}\frac{M}{0.5m_H}kT, \tag{5.14}$$

so

$$\frac{3}{2}\frac{M}{0.5m_H}kT = \left(\frac{M}{2m_H}4.5 \text{ eV} + \frac{M}{m_H}13.6 \text{ eV}\right). \tag{5.15}$$

The typical particle energy at the new equilibrium is then

$$\frac{3}{2}kT = \frac{1}{4}4.5 \text{ eV} + \frac{1}{2}13.6 \text{ eV} = 8.0 \text{ eV,} \tag{5.16}$$

and the temperature is

$$T = \frac{2}{3}\frac{8 \text{ eV}}{0.86 \times 10^{-4} \text{ eV K}^{-1}} = 60,000 \text{ K.} \tag{5.17}$$

To find the radius of the clump at this new equilibrium, we can again invoke the virial theorem,

$$\frac{1}{2}\frac{GM^2}{r} \sim \frac{3}{2}kT\frac{M}{0.5m_H}, \tag{5.18}$$

or

$$r \sim \frac{GMm_H}{6kT} = \frac{6.7 \times 10^{-8} \text{ cgs} \times 2 \times 10^{33} \text{ g} \times 1.7 \times 10^{-24} \text{ g}}{4 \times 8 \text{ eV} \times 1.6 \times 10^{-12} \text{ erg eV}^{-1}}$$

$$= 5.0 \times 10^{12} \text{ cm} \approx 0.3 \text{ AU.} \tag{5.19}$$

Thus, dissociation and ionization of hydrogen alone can consume enough energy to bring a solar-mass protostar down to dimensions of order 100 times that of the Sun. In reality, this adiabatic approximation for cloud collapse gives only an upper limit to the temperature and radius of the protostar. Several processes by means of which the gas can radiate away energy (and which are discussed in section 5.2.3, below) operate efficiently during the collapse, lowering the temperature considerably. Furthermore, as when discussing stellar structure, we have ignored angular momentum and magnetic fields, but in stellar formation they likely play important roles. In particular, angular momentum conservation will restrain the collapse in the plane perpendicular to the direction of rotation of the cloud, and an accretion disk structure (see discussion in section 4.6.1) will be formed.

From this point on, the loss of thermal energy is controlled by the radiative efficiency and the opacity of the gas, and the protostar at the center of the accretion disk slowly grows in mass by means of the matter channeled into it by the disk, and contracts in radius. Calculations indicate that after $\sim 10^7$–10^8 yr, the temperature of $\sim 10^7$ K required for hydrogen ignition is achieved in the core of the protostar, and it becomes a star on the main sequence in the H-R diagram. While there is a thus plausible physical path to star formation, it is important to realize that the processes described above have not been directly observed, since they occur inside dusty, and hence opaque, molecular clouds.

5.1.3 Planetary System Formation

The field of pre-main-sequence stellar evolution and planetary system formation is rich and rapidly developing, in terms of both theory and observations, but we will touch upon it only briefly and descriptively.

In recent years, disks of gas and dust have been discovered around many young stellar objects (see Fig. 5.1). As noted above, disks are expected naturally since the angular momentum of a collapsing gas clump prevents it from collapsing in a purely radial direction. These **protoplanetary disks** are thought to be potential planetary systems in the making. One possible scenario for planet formation is that the rocky cores of planets form first by mutual adhesion of colliding small fragments in the disk, followed by gravitational accretion of additional solid and gas-phase material. In an analogy of "the rich get richer," the superior gravitational attraction of the largest *protoplanets* allows them to accrete the most material from the disk and to swallow smaller planets, eventually concentrating much of the disk material in a fairly small number of planets. In the lower-temperature outer regions of the protoplanetary disk, beyond the *ice radius* (or *snow line*) where water is frozen, there is perhaps a larger quantity of solid material available for this process, compared to the hot inner regions of the disk, where water and CO_2 are in their gas phases. Icy planets and gas giants with large masses can thus form only beyond the ice radius.

In the final stages of pre-main-sequence contraction and initial nuclear ignition, a star goes through a *T-Tauri* phase, characterized by a strong stellar wind. Bipolar jets of material are also observed to emerge from some objects. When the stellar wind begins, it likely cleans out a star's surroundings of any remaining gas and dust that are not bound in planets. In the inner regions of the new planetary system, mainly the rocky cores of the planets remain. In the case of the Earth, the oceans and the atmosphere may have been later additions, brought to Earth via comet impacts (see Problem 1).

The recent discovery of extrasolar planets that are gas giants in small-radius orbits (see section 2.2.4) suggests a process of "migration" of the giant planets inward. It is presently still debated what is the physical process behind such migration, and why a migration of giant planets did not occur in our Solar System. It is unlikely that it did occur, all the way until the "swallowing" of one or more of the original giant planets by the Sun, since the migrating giant planets would have dragged the rocky inner planets along with them into the Sun.

5.1.4 The Initial Mass Function

Stars form with a range of masses between those of stars at the hydrogen-burning limit, $M \sim 0.1 M_\odot$, and the most massive known stars, $M \sim 100 M_\odot$. The relative number distribution of stars as a function of mass is described by the **initial mass function**. There is some evidence that this function is "universal," i.e., of the same form at all locations, and perhaps also at all times in the past, but there is yet no full agreement on this issue. Observations suggest that a reasonable description of the number of stars of mass m per unit mass interval is given by a power law,

$$\frac{dN}{dm} \propto m^{-\alpha}, \tag{5.20}$$

Figure 5.2 *Left:* The Pleiades, an open star cluster that is visible to the naked eye, at a distance of 120 pc. Note the nebulosities surrounding some of these young stars. Image scale is about 2 pc on the vertical side. *Right:* The globular cluster M80, at a distance of 8.7 kpc. Image scale is 4 pc on the vertical side. Photo credits: NASA, ESA, AURA/Caltech; NASA and the Hubble Heritage Team.

where $\alpha = 2.35$ over most of the mass range. This is called the Salpeter initial mass function. Thus, low-mass stars are much more common than high-mass stars. On the other hand, in the range between $0.1 M_\odot$ and $1 M_\odot$, the slope probably flattens, and there may even be a maximum to the distribution between $0.1 M_\odot$ and $0.5 M_\odot$. The faintness of low-mass stars, on the one hand, and the rarity and short lives of high-mass stars, on the other hand, make measurement of the initial mass function a difficult task. This empirical uncertainty compounds the already-mentioned theoretical uncertainty regarding the details of star formation.

5.1.5 Star Clusters

Due to the "top-down" fragmentation by which molecular clouds collapse, stars are formed in **star clusters**. Two main types of clusters are observed: open clusters and globular clusters. **Open clusters** are collections of 50–1000 stars, and are generally (but not always) observed to be young, as evidenced by the presence of massive stars. Most open clusters are not bound by their self-gravity, and therefore disperse on a **crossing timescale**, the time it takes a star to cross the cluster, given the typical random internal velocities. Since young open clusters are currently being formed from gas that has been enriched by previous stellar generations, the stars in open clusters are observed to have high "metallicities," i.e., "heavy" element abundances that are comparable to solar ones.

Globular clusters are bound systems of 10^4–10^6 stars within a spherical distribution of radius of a few parsecs. Their stellar populations consist of main-sequence stars and giants of masses lower than about a solar mass, as well as the stellar remnants (white dwarfs, neutron stars) of the more massive stars that existed in the past. The stars have low metallicities, indicating these are ancient systems that were formed from relatively unprocessed gas. Figure 5.2 shows examples of an open cluster and a globular cluster.

A third kind of cluster has been discovered in recent years in galactic regions called *starbursts* undergoing intense star formation. The clusters, coined **super star clusters**, consist of 10^4–10^6 stars within a few pc, including young, very massive stars. It is possible that super star clusters, or some fraction of them, can remain bound even after the massive stars in them explode as supernovae, and that they can survive other subsequent disruptive processes such as tidal forces from the galactic structures, which we will discuss in chapter 6. If so, after several billion years they will evolve into objects similar in many properties to globular clusters. Thus, super star clusters may simply be young globular clusters.

5.2 H II Regions

In star-birth regions, the more massive among the stars that have formed have high effective temperatures. These stars therefore emit UV and X-ray radiation, which ionizes the hydrogen and helium in the uncollapsed parts of the molecular cloud. The resulting **H II regions** consist of gas in equilibrium between **photoionization** by the stellar photons and the inverse process of **recombination** that occurs when ions capture free electrons. ("H II" is an astronomical term for ionized hydrogen, i.e., H^+. Thus, H I is neutral atomic hydrogen, C IV is three-times-ionized carbon, or C^{3+}, etc.) Let us examine some of the physical processes involved, and their consequences.

5.2.1 The Strömgren Radius

An approximately spherical region of radius r will be formed around a hot star, within which all the photons that can ionize hydrogen (photons with energies $h\nu > 13.6$ eV) are absorbed by the gas (see Fig. 5.3). Under steady-state conditions, the total number of hydrogen recombinations per unit time inside this volume must equal the total number of photoionizations per unit time. The latter rate equals just the number of ionizing photons that leave the star per unit time, Q_*. Thus,

$$Q_* = \mathcal{R}_{\text{rec}} \frac{4}{3} \pi r^3, \tag{5.21}$$

where $\mathcal{R}_{\text{rec}}$ is the number of recombinations of electrons and protons per unit time per unit volume. As in our discussion of nuclear reaction rates (see Eq. 3.123), recombination involves collisions between two types of particles having certain volume densities, relative velocities v, and a cross section σ_{rec} for interaction. Thus, the recombination rate will be

$$\mathcal{R}_{\text{rec}} = n_p n_e \langle \sigma_{\text{rec}} v \rangle = n_p n_e \alpha(T) = \alpha(T) n_e^2 = \alpha(T) x^2 n^2, \tag{5.22}$$

where

$$\alpha(T) \equiv \langle \sigma_{\text{rec}} v \rangle \tag{5.23}$$

is the temperature-dependent *recombination coefficient*. In Eq. 5.22, we have assumed that the density of electrons and protons is the same, $n_e = n_p$ (as it would be in a pure hydrogen gas), and that a fraction x of the gas, which has density n, is ionized. Let us further assume that the hydrogen is almost fully ionized within this region, i.e., $x \approx 1$, an assumption

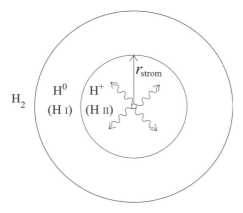

Figure 5.3 Schematic illustration of the *Strömgren sphere* of ionized hydrogen that is produced around a hot star via photoionization. Inside the radius r_{strom}, the hydrogen is almost completely ionized, i.e., there is an "H II" region. Beyond r_{strom}, the hydrogen is neutral ("H I"), since all photons capable of ionizing it have been absorbed. Further yet from the star, all photons energetic enough to photodissociate hydrogen molecules, H_2, have also been absorbed, and the gas is molecular.

we will justify shortly. Combining Eqs. 5.21 and 5.22, the radius of the spherical volume within which all the ionizing photons are absorbed is

$$r_{\text{strom}} = \left(\frac{3Q_*}{4\pi \alpha n^2} \right)^{1/3}. \tag{5.24}$$

This is called the **Strömgren radius**.

To obtain a typical value of r_{strom}, consider a main-sequence O5V star, with a luminosity per unit frequency L_ν. Its output of ionizing photons can be calculated numerically to be

$$Q_* = \int_{h\nu=13.6 \text{ eV}}^{\infty} \frac{L_\nu(\text{O5V})}{h\nu} d\nu \approx 3 \times 10^{49} \text{ s}^{-1}. \tag{5.25}$$

Typical gas densities in H II regions are $n = 10\text{--}10^4$ cm^{-3}. The recombination coefficient $\alpha(T)$ can be calculated from quantum mechanics. As will be elaborated in more detail below, the temperature of the gas is set by a balance between **heating** via photoionization and **cooling** through radiation. The equilibrium temperature is typically 10^4 K. At this temperature,[1]

$$\alpha(10^4 \text{ K}) = 2.6 \times 10^{-13} \text{ cm}^3 \text{ s}^{-1}. \tag{5.26}$$

---

[1] This value for $\alpha(T)$ includes recombinations of hydrogen to all its excited states, but excludes recombinations to the ground state. An atom that recombines to the ground state will emit a photon with energy $h\nu > 13.6$ eV, which will quickly ionize some other nearby hydrogen atom. Thus, recombinations to the ground state, under most conditions, have no effect on the overall ionization balance, and can therefore be ignored. This assumption is called "case B" recombination.

Thus, for example, for $n = 10^4$ cm^{-3},

$$r_{strom} = \left(\frac{3 \times 3 \times 10^{49} \text{ s}^{-1}}{4\pi \times 2.6 \times 10^{-13} \text{ cm}^3 \text{ s}^{-1} \times (10^4 \text{ cm}^{-3})^2} \right)^{1/3}$$

$$= 6 \times 10^{17} \text{ cm} = 0.2 \text{ pc}. \tag{5.27}$$

Such H II regions of ionized gas, which have been carved out of the molecular gas by the ionizing radiation around young stars are, in fact, seen in many star-forming regions (see Fig. 5.1).

5.2.2 Ionization Fraction

Outside the Strömgren radius the gas is, of course, neutral, since, by definition, the ionizing photons have all been absorbed. Let us now justify the assumption that, within the Strömgren sphere, the gas is almost completely ionized, and there is a sharp transition in hydrogen ionization—an **ionization front**—at r_{strom}. Using again the same considerations as before to write the rate of a reaction per unit volume, but now with collisions between neutral atoms and ionizing photons, the photoionization rate per unit volume at any point is

$$\mathcal{R}_{ion} = n_{photon} n(1 - x)\sigma_{ion} c, \tag{5.28}$$

where n_{photon} is the ionizing photon density, $n(1 - x)$ is the density of neutral hydrogen atoms, and σ_{ion} is the photoionization cross section. The photon density is just the ionizing photon flux at a distance r from the ionizing star, divided by c,

$$n_{photon} = \frac{Q_*}{4\pi r^2 c}. \tag{5.29}$$

The cross section of a hydrogen atom for photoionization from its ground state by photons with energies greater than $h\nu_0 = 13.6$ eV is

$$\sigma_{ion} \approx \sigma_0 \left(\frac{\nu}{\nu_0} \right)^{-3}, \quad \sigma_0 = 6.3 \times 10^{-18} \text{ cm}^2. \tag{5.30}$$

(Note, for comparison, that this cross section is 10^7 times larger than the Thomson cross section posed by a free electron. Thus, neutral hydrogen atoms are very easily ionized by photons with an energy of 13.6 eV.) Photons with energies $h\nu > 13.6$ eV will generally be in the exponentially falling Wien tail of the spectral distribution produced by the central star (see below). It is therefore safe to make the approximation that all of the ionizing-photon rate, Q_*, consists of photons with $h\nu \approx 13.6$ eV, and we can adopt σ_0 for σ_{ion}. Thus,

$$\mathcal{R}_{ion} = \frac{Q_*}{4\pi r^2} n(1 - x)\sigma_0. \tag{5.31}$$

In a steady state, at every point in the gas there will be a balance between the ionization rate per unit volume and the recombination rate per unit volume, $\mathcal{R}_{\rm ion} = \mathcal{R}_{\rm rec}$, or

$$\alpha(T)x^2 n^2 = \frac{Q_*}{4\pi r^2} n(1-x)\sigma_0. \tag{5.32}$$

This defines a quadratic equation for the ionization fraction x,

$$\frac{x^2}{1-x} = b, \quad b \equiv \frac{Q_* \sigma_0}{4\pi r^2 \alpha n}, \tag{5.33}$$

or

$$x^2 + bx - b, \tag{5.34}$$

with a positive solution (the negative solution is physically meaningless)

$$x = \tfrac{1}{2}\left(-b + \sqrt{b^2 + 4b}\right). \tag{5.35}$$

Let us estimate x at a typical location, say, $r = 0.1$ pc, inside the Strömgren radius of 0.2 pc that we found previously. For that particular choice of parameters,

$$
\begin{aligned}
b &= \frac{Q_* \sigma_0}{4\pi r^2 \alpha n} \\
&= \frac{3 \times 10^{49}\ {\rm s}^{-1} \times 6.3 \times 10^{-18}\ {\rm cm}^2}{4\pi\,(0.1 \times 3.1 \times 10^{18}\ {\rm cm})^2\ 2.6 \times 10^{-13}\ {\rm cm}^3\ {\rm s}^{-1}\ 10^4\ {\rm cm}^{-3}} = 6 \times 10^4,
\end{aligned} \tag{5.36}
$$

and

$$x = \tfrac{1}{2}\left(-b + \sqrt{b^2 + 4b}\right) = 0.999983, \quad 1 - x = 1.7 \times 10^{-5}. \tag{5.37}$$

Thus, we see that, indeed, the gas interior to the Strömgren radius is almost completely ionized.[2]

5.2.3 Gas Heating and Cooling Mechanisms

The rate at which energy is fed by the star into the H II region is called the **heating rate**. To calculate it, consider a free electron (called a *photoelectron*) that has just been liberated from

---

[2] In this derivation, we have assumed that the ionizing flux decreases only by geometrical dilution, as $4\pi r^2$. This assumption must obviously fail as we approach $r_{\rm strom}$, and the ionizing photons are depleted by the cumulative absorption of the small, but nonzero, neutral fraction of the gas between $r = 0$ and $r_{\rm strom}$. The thickness of the transition zone between ionized and neutral gas is of order the mean free path of an ionizing photon, $d \sim 1/n\sigma_0 = 5 \times 10^{-6}$ pc.

In contrast to the result above, a large neutral fraction within the Strömgren radius can arise in situations where the ionizing source has a "hard" spectrum, i.e., a large fraction of the ionizing photons have energies significantly larger than 13.6 eV. Because of the ν^{-3} dependence of the photoionization cross section, the ionization rate is then low, the ionization fraction x can be correspondingly low, and the transition zone is no longer small compared to $r_{\rm strom}$. Such conditions arise, e.g., in the gas that is photoionized by the X-ray photons from accreting objects such as X-ray binaries and active galactic nuclei (see chapters 4 and 6).

an atom by photoionization by an O star. Suppose the star has an effective temperature of $T_E = 20,000$ K. The peak of a blackbody distribution at this temperature is at

$$hv_{\max} = 2.8kT = 2.8 \times 0.86 \times 10^{-4} \text{ eV K}^{-1} \times 2 \times 10^4 \text{ K} = 4.8 \text{ eV}. \qquad (5.38)$$

Only photons with energy $hv > 13.6$ eV, i.e., in the Wien tail of the distribution, can ionize the hydrogen atoms, and all of these photons will be absorbed within the Strömgren sphere. The electrons liberated by the photoionization of hydrogen will have an energy equal to the ionizing photon energy, minus the 13.6-eV binding energy of atomic hydrogen. The total heating rate, integrated over the whole H II region volume, will therefore be

$$\Gamma_{\text{heating}}^{\text{tot}} = \int_{hv=13.6 \text{ eV}}^{\infty} (hv - 13.6 \text{ eV}) \, \frac{L_v(T = 20,000 \text{ K})}{hv} \, dv. \qquad (5.39)$$

The mean heating rate per unit volume is just $\Gamma_{\text{heating}}^{\text{tot}}$ divided by the volume of the Strömgren sphere. If the gas is dense enough, the photoelectrons quickly collide with the already-present electrons, protons, and ions in the gas and the photoelectron's kinetic energy is redistributed among all the particles, which follow a Maxwell-Boltzmann distribution with a well-defined temperature.

However, to photons, the rarified, ionized gas is largely transparent. The mean free path to Thomson scattering is

$$l = \frac{1}{n\sigma_T} = \frac{1}{10^4 \text{ cm}^{-3} \times 6.7 \times 10^{-25} \text{ cm}^2} = 1.5 \times 10^{20} \text{ cm} = 50 \text{ pc}, \qquad (5.40)$$

which is much larger than r_{strom}. Thus, most of the radiation escapes from the H II region without scattering. This means that in the H II region there may be a thermodynamic equilibrium between mass particles (which therefore follow a Maxwell-Boltzmann energy distribution), but not between mass particles and radiation. As a consequence, the radiation emitted by the 10^4-K gas will not have a blackbody spectrum.

Most of the energy deposited in the H II gas by ionizing photons is radiated away in the form of **emission line** photons at discrete wavelengths. To see how this comes about, let us return to the hydrogen atom. A recombining hydrogen atom will recombine either to the ground state or to some excited state, while radiating its excess energy as a "continuum" photon of energy equal to the difference between the kinetic energy of the free electron and energy of the atomic state (see Fig. 5.4). The atom will then decay radiatively in one or more steps until it reaches the ground state, while emitting an emission line photon at every step, with an energy corresponding to the difference between the atomic energy levels. The probabilities for the atom to recombine into each of the states, and then to decay into each of the lower states, can again be calculated from quantum mechanics. The last decay usually involves emission of a Lyman series photon (i.e., a transition to the ground level).

The hydrogen inside the Strömgren radius is constantly being ionized and recombining, and therefore at a given moment, some small fraction of it is in atoms in the ground

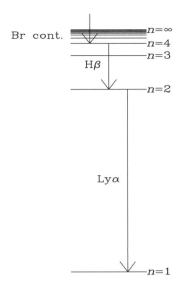

Figure 5.4 Example of a recombination cascade in a hydrogen atom. A free electron and proton recombine into the $n = 4$ excited level of hydrogen, emitting the sum of the kinetic energy of the free electron and the binding energy of the $n = 4$ level, in the form of a *Brackett continuum* photon. The atom then decays radiatively to the $n = 2$ level, emitting an Hβ (Balmer series) emission-line photon. Finally, the atom decays to the ground level and emits a Lyman-α photon.

state. As already noted, hydrogen in its ground state has a very high cross section for absorbing Lyman photons and, as a result, the emitted Lyman photon is almost immediately reabsorbed by some other atom, which again decays in one or more steps ending in the emission of a Lyman photon. This process is repeated until the excitation energy of the original state, into which an atom recombined, has been degraded into many lower-energy hydrogen transition photons (Balmer, Paschen, Brackett, etc.) and one Lyman-α photon. The lower-energy photons escape readily from the gas, while the Lyman-α photons execute a random walk of absorption and reemission until reaching the edge of the H II region. If there is dust mixed with the gas (more on dust soon), a Lyman-α photon may be absorbed by a dust grain before it reaches the edge of the ionized region, in which case the photon energy is reemitted by the grain as infrared thermal radiation. The escape of photons of this hydrogen *recombination emission* is one way in which the ionized gas "cools," i.e., loses the energy that is constantly being fed into it by the central star.[3]

---

[3] Strictly speaking, recombination *lines* are *not* a gas cooling mechanism. By means of the recombination lines, the gas gets rid of the binding energy of its constituent atoms, but not of its heat (i.e., the kinetic energy of its constituents). The emitted energy of recombination line photons is always less than the 13.6 eV of ionization energy per atom that we subtracted in Eq. 5.39, and which therefore does not contribute to the gas temperature. Only the hydrogen *continuum* photons (Balmer continuum, Paschen continuum, etc.), minus the binding energy of their appropriate lower energy levels, contribute to gas cooling via hydrogen recombination emission.

The luminosity emitted per unit volume in a particular hydrogen emission line, say, Hα ($n = 3 \rightarrow 2$ in the Balmer series), is (using Eq. 5.22)

$$\mathcal{L}_{H\alpha} = \alpha_{H\alpha}(T)n_e^2 h\nu_{H\alpha}, \tag{5.41}$$

where $\alpha_{H\alpha}(T)$ is the effective recombination coefficient for hydrogen recombinations that lead to the emission of Hα. A similar equation can be written for, say, Balmer continuum photons, a fraction of whose energies actually cools the gas. Thus, the rate of energy loss, or **cooling rate** of the gas, due to hydrogen recombination is proportional to the gas density squared. Note that, since most recombinations end with the emission of one Lyman-α photon, and the rate of recombinations equals the rate at which the central star emits ionizing photons, the Lyman-α photon emission rate approximates the ionizing photon rate. Thus, measuring the Lyman-α flux is a practical way of deducing the ionizing flux of the star (i.e., the flux with $h\nu > 13.6$ eV).[4]

A second cooling mechanism, which is much more efficient than hydrogen recombination, is by means of line emission from ions of heavier elements. The electrons and ions (mostly protons) in the gas collide with the ions of various elements, most importantly carbon, nitrogen, oxygen, sulfur, silicon, and iron (so-called "metals"), that exist in the gas in trace amounts. These ions have excited energy levels at relatively low energies above the ground level, similar to the typical kinetic energies of the electrons. An electron–ion collision can therefore lead to **collisional excitation** of the ion, which takes up some of the kinetic energy of the electron. If another collision does not *deexcite* the excited ion and thus returns the energy to the gas, the excited ion will decay radiatively to its lower energy levels, and the resulting emission-line photons will carry away the energy—there is a low probability for reabsorption of the photons by other ions, because their abundance is so low.

Even if collisional deexcitation does occur much faster than radiative decay, this cooling mechanism still operates. Consider a whole population of a particular ion in a gas where, as before, there is thermodynamic equilibrium between mass particles, but not necessarily between mass particles and radiation.[5] The ratio between the population of any two of the ion's energy levels is then, by definition, given by the **Boltzmann factor**,

$$\frac{n_2}{n_1} = \frac{g_2}{g_1}e^{-h\nu/kT}, \tag{5.42}$$

where $h\nu$ is the energy difference between the levels, and g_1 and g_2 are the statistical weights of the levels. Thus, as long as $h\nu$ is not much greater than kT, there will be an equilibrium population of ions in the excited level, and the constantly ongoing radiative decay of that population will lead to the loss of thermal energy by the gas, in the form of emission line photons from the ions.

---

[4] Under low-density ($\ll 10^4$ cm^{-3}) conditions, about one-third of the atoms in the $n = 2$ state can decay to the ground state via *two-photon decay* instead of through Lyman-α emission. The two photons have a total energy that equals the energy of a Lyman-α photon [13.6 eV$(1 - 1/4) = 10.2$ eV], but with individual energies distributed in a continuum between 0 and 10.2 eV and peaked at 10.2 eV/2 = 5.1 eV.

[5] In the context of the physics of diffuse interstellar gas, this situation is called **local thermodynamic equilibrium**. Note that in other contexts, this term sometimes has a different definition. For example, in the context of stellar structure, local thermodynamic equilibrium refers to a situation in which the temperature may vary with location in the star, yet at a particular position there is thermodynamic equilibrium, including a radiation field with a Planck spectrum that corresponds to the local kinetic (Maxwellian) temperature.

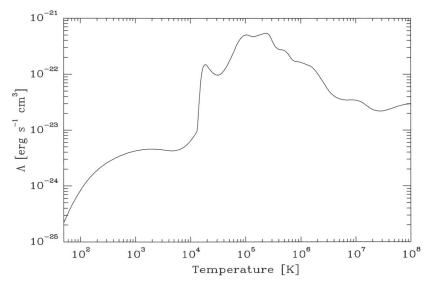

Figure 5.5 Example of a cooling function for interstellar gas, under the assumption that the ionization is purely collisional at every temperature (i.e., there is no photoionization). The complex shape of the function is due to the different cooling mechanisms that are dominant at different temperatures: molecular line emission at $T < 10^3$ K; at $T \sim 10^3$–10^4 K, collisionally excited lines of neutral and low-ionization species of metals; at $T \sim 10^4$–10^5 K, hydrogen and helium recombination; at $T \approx 10^5$ K, $T \approx 2 \times 10^5$ K, $T \approx 5 \times 10^5$ K, and $T \approx 10^6$ K, far-UV and X-ray emission lines from highly ionized species of carbon, oxygen, neon, and iron, respectively; and bremsstrahlung at $T > 10^7$ K. Note that heating sources (e.g., stars) will affect the ionization levels of the different elements, and will therefore influence the form of the cooling function.

A third type of radiative cooling process, which is efficient mainly under high-density conditions, is bremsstrahlung (free–free) emission. Bremsstrahlung can also become the dominant (even if inefficient) cooling mechanism in gas that is both sufficiently hot (hence fully ionized) and tenuous (so electrons and ions rarely meet and recombine, and therefore the recombination rate is low—see Eq. 5.22). Such conditions exist in the *intracluster medium* of galaxy clusters (see section 6.4).

The total cooling rate of an interstellar gas will thus be a sum of the cooling rates due to all the various cooling processes. Which processes are dominant will depend on the gas composition (chemical abundances and ionization fractions), the density, and the temperature. Figure 5.5 shows an example of a *cooling function*, $\Lambda(T)$, of interstellar gas. The cooling rate per unit volume can be obtained by multiplying the cooling function, which has units of power times volume, by the gas density squared. The complex form of the curve is due to the changing dominance of different cooling processes at different temperatures. Note, in particular, that the cooling function reaches a maximum at $T \sim 10^5$ K. Beyond $T \sim 10^7$ K, all the elements in the gas are almost fully ionized and bremsstrahlung remains the only, relatively inefficient, cooling avenue. Because the cooling rate, $\Lambda(T)n^2$, is very low for a gas that is both hot and tenuous, such a gas may remain "stuck" at a high temperature for a very long time.

The efficacy of a cooling process may also depend on the opacity of the gas to radiation at the wavelengths of the cooling emission—if the gas is opaque, the energy cannot escape. As in the interiors of stars, the opacity depends on chemical composition, density, and temperature. The temperature, in turn, is determined under steady-state conditions by the requirement that the heating rate per volume element equal the cooling rate per volume element,

$$\Gamma_{\text{heating}} = \Gamma_{\text{cooling}}. \tag{5.43}$$

To examine more quantitatively the relative importance of the various gas cooling processes under different conditions, consider a population of hypothetical two-level ions that can radiate emission-line photons of energy $h\nu$. The ions are constantly being excited by collisions with electrons, and deexcited by both collisional and radiative transitions. The density of electrons is n_e, and the densities of ions in level 1 and in level 2 are n_1 and n_2, respectively. The total density of these ions is $n = n_1 + n_2$. Some fraction of the gas is in the ions, as determined by the balance between ionization due to the radiation field and recombination with electrons, and hence $n \propto n_H \propto n_e$. The rate of collisional excitations from level 1 to level 2 of the ion is proportional to the electron density and to the density of ions in level 1. Thus,

$$\mathcal{R}_{12,\text{coll}} = n_e n_1 q_{12} = n_e (n - n_2) q_{12}, \tag{5.44}$$

where q_{12} is a collisional excitation coefficient that includes the effects of cross section and temperature. Similarly, the rate of collisional deexcitations from level 2 to level 1 is

$$\mathcal{R}_{21,\text{coll}} = n_e n_2 q_{21}, \tag{5.45}$$

where q_{21} is the corresponding collisional deexcitation coefficient. The rate of radiative decay between the two levels is

$$\mathcal{R}_{21,\text{rad}} = n_2 A_{21}, \tag{5.46}$$

where A_{21} is the Einstein coefficient for spontaneous radiative emission, i.e., the reciprocal of the mean lifetime for spontaneous decay to the lower energy level. (We will here ignore stimulated emission, which is unimportant in most circumstances, but, for an exception, see the discussion of masers in the next section.) In equilibrium, the rate of excitations must equal the rate of deexcitations, i.e.,

$$n_e(n - n_2)q_{12} = n_2 n_e q_{21} + n_2 A_{21}. \tag{5.47}$$

Isolating n_2 gives

$$n_2 = \frac{n_e n q_{12}}{n_e(q_{21} + q_{12}) + A_{21}}. \tag{5.48}$$

The luminosity in emission-line photons per unit volume is just

$$\mathcal{L}_{21} = n_2 A_{21} h\nu = \frac{n_e n q_{12} A_{21} h\nu}{n_e(q_{21} + q_{12}) + A_{21}} = \frac{n_e n q_{12} h\nu}{(1 + q_{12}/q_{21})n_e q_{21}/A_{21} + 1}$$

$$= \frac{n_e n q_{12} h\nu}{(1 + q_{12}/q_{21})n_e/n_{\text{crit}} + 1}, \tag{5.49}$$

where we have defined a **critical density**, $n_{\text{crit}} \equiv A_{21}/q_{21}$. The ratio between the collisional excitation and deexcitation coefficients is related by the statistical weights, g_1 and g_2, of the levels and by the Boltzmann factor[6],

$$\frac{q_{12}}{q_{21}} = \frac{g_2}{g_1} e^{-h\nu/kT}, \tag{5.50}$$

and therefore for any temperature, this ratio is of order 1 or less. Looking at the right-hand term in Eq. 5.49 we see that, if $n_e \ll n_{\text{crit}}$, the first term in the denominator can be neglected, and the luminosity depends on electron density as

$$\mathcal{L}_{21} \sim n_e n \propto n_e^2. \tag{5.51}$$

However, when $n_e \gg n_{\text{crit}}$, the second term in the denominator can be neglected and then

$$\mathcal{L}_{21} \sim n \propto n_e, \tag{5.52}$$

i.e., the luminosity grows only linearly with density. Since cooling by hydrogen recombination and by bremsstrahlung is proportional to n_e^2 at all densities, cooling by heavy-element-ion line emission becomes inefficient compared to cooling by hydrogen when the density is above the critical density of the major cooling lines.

A good example of a collisional-excitation line coolant is the [O III]$\lambda\lambda$ 4959, 5007 doublet, emitted by doubly ionized oxygen, which is one of the main coolants of H II regions. (The square brackets indicate that these are "forbidden" quantum transitions, which have low probability and hence a long lifetime for radiative decay.) It is emitted by the transition of an O^{2+} ion from an excited level to the fine-split ground level (see Fig. 5.6). At densities lower than its critical density for collisional deexcitation, $n \sim 10^6$ cm^{-3}, a collisionally excited O^{2+} ion has enough time between collisions to release its energy in photons of these two wavelengths. These photons give H II regions (e.g., the Orion nebula), when viewed through the eyepieces of small telescopes, their greenish hue. When these emission lines were first discovered in the 19th century, they could not be identified with any known atomic transition, and it was thought that they came from a new, unknown element, dubbed "nebulium." The [O III]$\lambda\lambda$ 4959, 5007 doublet is difficult to observe in the laboratory because, even in the best achievable laboratory vacua, the density is higher than the critical $n \sim 10^6$ cm^{-3}. At the low gas densities of H II regions, the [O III]$\lambda\lambda$ 4959, 5007 doublet is one of the primary avenues by which the gas gets rid of the energy being deposited into it by the central star via photoionization.

Remarkably, not only are the rare "metals" the main coolants of H II regions, but they also serve as natural thermostats that keep H II gas always at about 10^4 K, regardless of the effective temperature of the ionizing star. If the star is hot and deposits a lot of energy into the gas, this raises the collisional excitation rate of the metals, and hence the rate at which they radiate away the energy in emission lines, preventing a rise in the temperature. If the star is relatively cool, there will be less collisional excitation of the metal ions, and hence

---

[6] This is a consequence of the fact that, if collisions dominate the level populations, then in equilibrium we have $n_1 q_{12} = n_2 q_{21}$. However, under such conditions Eq. 5.42 holds (by definition), and the stated result then follows.

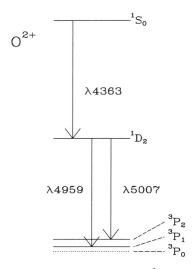

Figure 5.6 Part of the energy-level structure of the O^{2+} ion, an important cooling agent of interstellar gas. Radiative decay from the $^1D_2$ state to two of the sublevels of the triply split ground state produces the emission-line doublet [O III] $\lambda\lambda 4959, 5007$. (Decay to the third sublevel, $^3P_0$, is strongly forbidden.) For clarity, the spacing between the split levels has been greatly exaggerated. The [O III] $\lambda 4363$ emission line, resulting from radiative decay from $^1S_0$ to $^1D_2$, can become strong in gas with a temperature high enough to excite a significant fraction of the ions to the $^1S_0$ state.

less loss of energy via emission lines. The gas temperature will then rise up to about 10^4 K, at which point heating and cooling will balance out. This equilibrium temperature is set mainly by the atomic properties of the metal coolants.

Figure 5.7 shows a typical optical spectrum of an H II region. A qualitatively similar spectrum is also emitted by a planetary nebula, where the ionizing source is a newly forming white dwarf, instead of a massive young star, and the ionized gas is the ejected stellar envelope from the red-giant phase, rather than the star-forming cloud. A detailed analysis of the relative and absolute strengths of the emission lines in such spectra can reveal a wealth of information about the physical conditions—the temperature, the density, and the chemical abundances of the gas, as well as the form of the ionizing spectrum of the source. For example, the O^{2+} ion emits not only the [O III]$\lambda\lambda$ 4959, 5007 doublet, by transition from level $^1D_2$ to the fine-split ground level, but also the singlet [O III]λ 4363 by decaying from level $^1S_0$ to level $^1D_2$ (see Fig. 5.6). The flux ratio of $f(4363)$ to $f(4959) + f(5007)$ provides a direct probe of the excitation level of this ion, which depends mainly on the gas temperature. Measurement of this ratio is therefore a useful thermometer.

Note that, although we have considered only heating by stellar radiation, the interstellar gas can be heated by other processes, including shock waves from stellar winds and supernovae, energetic charged particles called *cosmic rays*, and various "background" radiation fields that are not associated with a particular source. Some of these are discussed in later chapters.

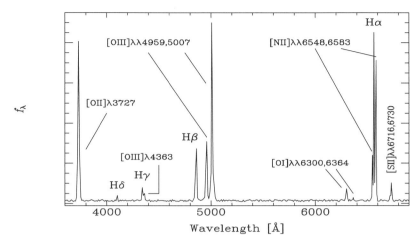

Figure 5.7 Typical optical spectrum of an H II region, with the main emission lines marked—the Balmer series from hydrogen recombination, and the collisionally excited "metal" lines.

5.3 Components of the Interstellar Medium

Beyond the Strömgren sphere of H II gas, photons from the ionizing source with energies $11.1 < h\nu < 13.6$ eV can **photodissociate** molecular hydrogen[7] into neutral hydrogen atoms, i.e., H I. At some larger radius, all of these photons will also be used up (or absorbed by interstellar dust—see below), and the molecular gas will therefore be fully shielded from the stellar radiation. In reality, star-forming regions do not have perfect spherical geometries because the clouds have irregular shapes, inhomogeneous densities, and unevenly distributed ionizing stars and star clusters within them.

An observationally important emission line from the neutral H I gas is the *hyperfine splitting* 21-cm radio emission line, at a frequency of 1.421 GHz. The ground state of hydrogen is split in energy due to the spin–spin interaction between the electron and the proton (see Fig. 5.8). The upper level corresponds to parallel spins (total angular momentum $F = 1$) and the lower level to antiparallel spins ($F = 0$). The upper level can be populated by collisions with other atoms, by absorption of photons of this wavelength from *background radiation* (the cosmic microwave background, to be discussed in chapter 9, is one such background), and by previous decays of the atom to the ground level via one of the Lyman transitions, which populate both hyperfine levels according to their statistical weights (3 to 1). Radiative decay from the $F = 1$ to the $F = 0$ level is "highly forbidden." The timescale is therefore 10^7 yr for the electron to do such a "spin-flip,"

---

[7] Although the binding energy of molecular hydrogen is only 4.5 eV, the cross section for direct photodissociation from the ground state (e.g., by absorption of an $h\nu > 4.5$ eV photon) is very low. Instead, molecular hydrogen photodissociates in a two-step process, first by radiative absorption of an $h\nu > 11.1$ eV photon to an excited electronic energy level. Part of the excited electron state's energy then unbinds the molecule, and the rest of the energy goes to continuum radiation and kinetic energy of the atomic fragments.

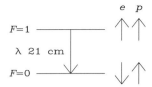

Figure 5.8 The hyperfine-split ground state of hydrogen. The lower level has angular momentum $F = 0$, corresponding to antiparallel electron and proton spins, and the upper level has $F = 1$, with the two spins parallel. In a radiative decay between the levels, the spin of the electron flips, and a $\lambda = 21$ cm ($\nu = 1421$ MHz, $h\nu = 5.9 \times 10^{-6}$ eV) photon is emitted.

with the accompanying spontaneous emission of a 21-cm photon. The long timescale means that the 21-cm line does not play a significant role in gas cooling.[8] Nevertheless, the large amounts of atomic hydrogen in many astronomical settings, and the relative ease of sensitive radio observations, make the 21-cm emission a bright and useful tracer of this phase of the interstellar medium.

The molecular-gas phase of the interstellar medium radiates primarily in transitions at IR through radio wavelengths. The main constituent of this gas, the H_2 molecule, is an inefficient radiator and its emission is therefore quite weak. In its place, the molecular phase can be traced in emission (or absorption of a background source) by the next most common molecules: CO (carbon monoxide), OH (hydroxyl), NH_3 (ammonia), and others. Although CO is only $\sim 10^{-5}$ as abundant as H_2, in many cases it is the dominant coolant of the molecular gas, and thus determines the temperature of the molecular gas phase. As was the case for atomic lines, analysis of the fluxes in different lines and their ratios provides a diagnostic of physical conditions in the gas (temperature, density, abundances) and of the embedded stars. Apart from the molecules already mentioned, hundreds of different types of molecules have been identified in the ISM by their spectra. Some of these molecules are fairly complex, for example, CH_3CH_2OH (ethanol—drinking alcohol), which consists of nine atoms.

The complex energy-level structures of molecules allow, in some molecules, the creation of a *population inversion*, in which a meta-stable level (i.e., one having a long lifetime for spontaneous radiative decay) is populated in many molecules. Given the appropriate gas geometry, radiative decay by **stimulated emission** from this level to a lower energy level can produce amplification of the radiation corresponding to the transition. The amplification grows exponentially with the path length. Many such astronomical **masers** (microwave amplification of stimulated emission radiation) have been discovered for transitions in OH, H_2O, and CH_3OH (methanol—"wood alcohol"). Observationally, one sees a compact region in a molecular cloud radiating in the masing transition with a huge intensity. Astronomical masers achieve the large path lengths required for amplification by means

---

[8] The main coolant of H I gas is usually a far-infrared line of singly ionized carbon, C II $\lambda 158 \mu$m, that is collisionally excited by impacts with neutral atoms. Warmer ($\sim 10^4$ K) neutral gas is cooled primarily by Lyman-α emission from collisionally excited hydrogen atoms.

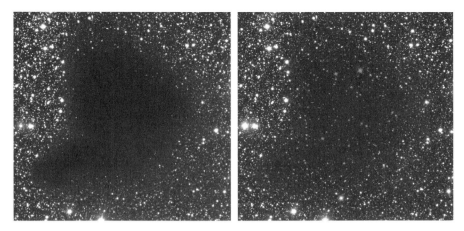

Figure 5.9 Images of the "dark cloud" Barnard 68 in optical light (left) and in the near infrared (right). The dust in a molecular cloud on the line of sight obscures the optical light from the gas and the stars behind it, while the IR light suffers less extinction by the dust. The cloud is at a distance of 120 pc, and is about 0.1 pc across. Photo credits: J. F. Alves et al., and the European Southern Observatory.

of the large distances available in astronomical settings. This is in contrast to lasers and masers in the laboratory, which generally achieve large path lengths by means of multiple reflections between mirrors.

The last important component of the ISM, to which we have already alluded several times, is **interstellar dust**. Dust consists of solid grains containing mainly Fe, Si, C, H_2O ice, and CO_2 ice, with a range of grain sizes, but typically $r \sim 0.001$–$0.1\,\mu$m. Given these grain sizes, a more appropriate name would be "interstellar smoke"—the particles of familiar tabletop dust are much larger. The formation and detailed composition of interstellar dust are not fully known yet. Dust is produced in the photospheres of evolved stars and in supernovae, and is expelled by stellar winds and by supernova explosions. Its presence is detected in astronomical observations either by the attenuation it causes in background light sources via scattering and absorption, or by the reemission of the absorbed energy as thermal infrared radiation. Attenuation of background light by dust is called interstellar **extinction**. In images of star-forming regions, dust extinction produces dark patches that obscure stars and fluorescing gas (see Fig. 5.9). As seen in Fig. 5.10, extinction increases steeply with decreasing wavelength in the IR through UV range. As a result, the short-wavelength emission from a source will be attenuated more than the long-wavelength emission, and dust will also cause the so-called **reddening** of background sources of light. Over most of the wavelength range shown, the extinction behaves roughly with an exponential dependence on wavelength, as $[\log E(\lambda)] \sim [\log E(\lambda_0)]\lambda_0/\lambda$. The "bump" at ≈ 2200 Å is likely due to absorption by a graphite-like component in the dust particles. Comparison of the observed color and brightness of a light source to its intrinsic color and brightness (as deduced, e.g., if the source is a star, from its absorption line spectrum) allows deducing the amount and properties of the dust along the line of sight. Apart from their observational effects, dust grains play an important physical role as sites to which atoms, ions,

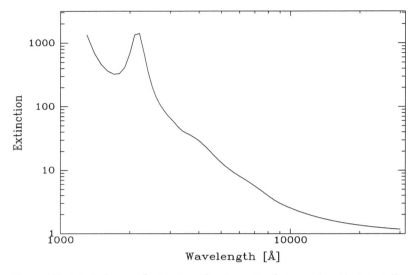

Figure 5.10 A typical curve of extinction—the attenuation factor produced by interstellar dust—as a function of wavelength, from the UV (1300 Å) to the near-IR (3 μm). The case shown is set to have a factor of 10 extinction at 5500 Å. Note the large extinctions in the UV, compared to the moderate values at IR wavelengths. Data credit: J. Cardelli et al. 1988, *Astrophys. J.*, 329, L33.

and simple molecules can temporarily attach, interact with other attached particles to form more complex molecules, and then return to the ISM. Furthermore, electrons released from dust grains by UV photons via the photoelectric effect are the main heating source of atomic gas.

5.4 Dynamics of Star-Forming Regions

In the preceding discussion of the ISM and H II regions, we have repeatedly invoked steady-state conditions. However, in some situations, dynamical evolution needs to be taken into account. When a new massive star "turns on," the ionization front it produces in the surrounding gas advances (relatively slowly; see Problem 4), from zero radius until equilibrium is reached at the Strömgren radius. Stars, both in their pre-main-sequence stage, and after they evolve off the main-sequence, when they are giants or supergiants, shed their outer envelopes and drive winds and dust into the surrounding ISM. These winds drive shocks into the ISM, affecting its temperature and density structure, as well as enriching the ISM with newly synthesized heavy elements.

The massive stars that end their evolution as core-collapse supernovae release in the explosion of order $1M_\odot$ or more of highly enriched *ejecta*, which plow through the previously shed stellar envelopes and through the ISM, at supersonic velocities of order 10^3–10^4 km s^{-1}. The ejecta drive strong shocks, which heat the ejecta and the surrounding material to keV temperatures, making supernova remnants bright X-ray sources (even when the remnant is no longer powered by the rotational energy of the neutron star, as in the case of the very young Crab nebula). At the shocks, electrons are accelerated to relativistic energies and spiral around magnetic field lines, emitting synchrotron radiation from radio to optical bands. Finally, the shocks from supernovae may initiate compression of the gas in neighboring molecular clouds, and thus set off new generations of star formation.

In any interstellar environment, many of the numerous processes we have discussed in this chapter, both in and out of equilibrium, play a role and are interdependent. As a result, the prediction of observables that can be compared to actual measurements, such as emission-line luminosities and ratios, generally requires numerical calculations that take into account all the relevant processes. The physical conditions in the ISM—temperature, density, chemical abundances, and heating sources—can then be deduced by comparing the measurements to the results of model calculations using grids of these physical input parameters. Knowledge of the physical conditions in different environments and at different stages can, in turn, lead to further understanding of star formation.

Problems

1. The oceans on Earth have a mean depth of 3.7 km and cover 71% of the Earth's surface. It has been suggested that this water was brought to Earth by comets (which are composed mainly of frozen water and CO_2).

   a. Calculate the kinetic energy of a spherical comet of radius 4 km, composed of water ice, which arrives from far away to the region of the Earth's orbit around the Sun.

   b. Estimate the radius of the cylindrically shaped crater that such a comet creates when it strikes the Moon. Assume that the crater, of depth 10 km, is formed by heating to 3500 K, and thus vaporizing, a cylindrical volume of moon rocks. Moon rocks are made of silicates, which have molecular weights around 30 (i.e., a typical molecule has 30 times the mass of a hydrogen atom), and mean solid densities $\rho \sim 2$ g cm^{-3}. Ignore the latent heat required to melt and vaporize the rocks, and the energy involved in vaporizing the comet itself.

   Hint: Equate the kinetic energy of the comet with the thermal energy of the vaporized rocks.

   Answer: 51 km.

   c. The number of craters per unit area in the relatively smooth "mare" regions of the Moon, which trace the impact history over the past $\sim$3 Gyr, indicate a total of about 10 impacts, leaving 50-km-radius craters, during this period. Based on the assumptions in (b), these would be impacts of objects with radii > 4 km. From geometrical considerations alone (i.e., the relative target sizes posed by the Earth and by the Moon, and ignoring gravity) estimate how many such objects have struck the Earth, and what is the mean time interval between impacts. How does the interval you found compare to $\sim$60 Myr, the typical interval between large extinctions of species on Earth? (The most recent large extinction, 65 Myr ago, eliminated the dinosaurs, and marked the rise of the mammals.)

Answer: 140 Earth impacts, 21 Myr mean interval. The impact rate was higher in the past, and declined as the comet reservoir was depleted, mainly by captures on Jupiter and the Sun.

d. Assume that comets have a mass distribution $dN/dm \propto m^{-3}$, with radii ranging from 0.2 to 4 km. Based on the number of 4-km comet impacts, show that the total comet mass, if composed mainly of frozen water, is sufficient to make Earth's oceans.

2. Consider a newly formed globular cluster, with a total mass $10^6 M_\odot$, and an initial mass function $dN/dm = am^{-2.35}$ in the mass range 0.1–20$M_\odot$, where $m \equiv M/M_\odot$.

   a. Find the constant a.

   Answer: 1.9×10^5.

   b. Find the total luminosity of the cluster, assuming that all its stars are on the main sequence, and a mass–luminosity relation $L \sim M^4$. What fraction of the luminosity is contributed by stars more massive than $5M_\odot$?

   Answers: $2.0 \times 10^8 L_\odot$; 0.98.

   c. Find the mean mass of a star in the cluster.

   Answer: $0.33 M_\odot$.

   d. Assume that the main sequence lifetime of a $1M_\odot$ star is 10 Gyr, and main sequence lifetime scales with mass as M^{-2}. What is the mass of the most massive main-sequence stars in the cluster after 1 Gyr? What is the total luminosity of the cluster at that time?

   Answers: $3.2M_\odot$; $1.5 \times 10^6 L_\odot$.

3. Assume that the Milky Way, the galaxy in which we live, is composed of $5 \times 10^{10} M_\odot$ of gas, and $\sim 10^{11}$ stars, which were formed with an initial mass function $dN/dM \propto M^{-2.35}$ in the range 0.4–100$M_\odot$.

   a. What fraction of the stars formed with a mass above $8M_\odot$, the lower limit for eventual core collapse? How many neutron stars and black holes are there in the galaxy, and roughly how much mass is there in these remnants?

   Answers: 0.018; 1.8×10^9 remnants; and $\sim 3 \times 10^9 M_\odot$, assuming a typical remnant mass of $1.4M_\odot$.

   b. Assume that every stellar core collapse, and the supernova explosion that follows it, distribute $0.05M_\odot$ of iron into the interstellar medium. What is the mean interstellar mass abundance of iron in the Galaxy? Compare your answer to the measured mass abundance of iron in the Sun, $Z_{Fe,\odot} = 0.00177$, and explain how this shows that the Sun is a "second-generation" star, that was formed from preenriched interstellar material.

   c. Several systems of "binary pulsars" are known, consisting of two neutron stars in close orbits. If half of all stars are in binaries, and members of binaries are formed by a "random draw" from the initial mass function (i.e., $P(m) \propto m^{-2.35}$), then how many pairs of stars in the Milky Way were formed in which *both* companions were more massive than $8M_\odot$?

Answer: 7.7×10^6.

d. Due to asymmetries in the supernova explosion, neutron stars are born with a "kick" that gives them a typical velocity of 500 km s^{-1}. What is the maximal initial separation that will allow a binary to remain bound?

Hint: Equate the binding energy between two $8M_\odot$ stars at separation r to the kinetic energy of two $1.4M_\odot$ neutron stars, each having typical kick velocity.

e. If binaries form with an initial separation distributed uniformly between 0 and 0.01 pc, how many neutron stars binaries have survived the formation kick?

Answer: 600.

Note: Actual measurements indicate that the secondaries in binaries are drawn not from a Salpeter mass function, but rather from a mass distribution that is approximately flat. Also, the measured binary separation distribution is not flat, as assumed in this problem, but rather flat in logarithmic intervals (or, equivalently, $dN/da \propto 1/a$, where a is the separation).

4. A new star lights up inside a cloud of atomic hydrogen with a constant number density of n atoms per unit volume. The star emits ionizing photons at a rate of Q_* photons per unit time. The ionizing photons begin carving out a growing "Strömgren sphere" of ionized gas inside the neutral gas.

a. At a distance r from the star, what is the timescale τ_{ion} over which an individual atom gets ionized, if the ionization cross section is σ_{ion}?

b. If the recombination coefficient is $\alpha \equiv \langle \sigma v \rangle$, what is the timescale τ_{rec} for an individual proton to recombine with an electron?

c. At a position close to the star, where the ionizing flux is high, and therefore $\tau_{ion} \ll \tau_{rec}$, show that the velocity at which the ionization front that bounds the Strömgren sphere advances is $v_{if} = Q_*/(4\pi r^2 n)$.

Hint: Assume, as usual, that the gas is completely ionized within the front, and completely neutral beyond it. Consider a slab of neutral gas behind the ionization front, with area ΔA and thickness Δr, and find the volume of neutral gas that is ionized during an interval Δt.

d. Evaluate v_{if} for $Q_* = 3 \times 10^{49}$ s^{-1}, $n = 10^4$ cm^{-3}, and for $r = 0.01$ pc, 0.05 pc , and 0.1 pc, respectively. From $v_{if}(r)$, obtain and solve a simple differential equation for $r_{strom}(t)$, and find roughly how long it takes the ionization front to reach the final Strömgren radius (0.2 pc for these parameters; see Eq. 5.27).

Answers: At $r = 0.01$ pc, $v_{if} = c$; at 0.05 pc, $v_{if} = 0.3c$ and at 0.1 pc, 2.5×10^4 km s^{-1}; about 10 years to reach r_{strom}.

6 | The Milky Way and Other Galaxies

Galaxies are concentrations of 10^7 to 10^{11} stars, and are essentially the only places where stars exist.[1] The Sun is a star in the Milky Way, a large **spiral galaxy**. Stars are, however, just one of several components that make up a galaxy. In this chapter, we examine the properties of the Milky Way, of other galaxies, and of systems of galaxies.

6.1 Structure of the Milky Way

Figure 6.1 shows optical-band images of spiral galaxies from different perspectives. Figure 6.2 shows schematically the various components of the Milky Way and of other spiral galaxies. The components, which we will describe in more detail below, are a flattened circular **stellar disk**, which has spiral-shaped density enhancements in it; a thinner, **gas-and-dust disk**, within the stellar disk; a stellar **bulge** or **spheroid**; an extended, spheroidal, **halo** of gas, stars, and **globular clusters**; a central **supermassive black hole**; **cosmic rays**— energetic protons, nuclei and electrons that are trapped by the magnetic fields in the disk; and a **dark halo** of unknown composition, which extends far beyond the visible components.

The Sun is in the disk of the Milky Way (often referred to as "the Galaxy," with a capital G). The diffuse band of light known by this name, and visible in the night sky, is composed of individual stars. Its appearance is the result of our vantage point inside the plane of the disk (see Fig. 6.3). The Sun's distance to the Galactic center is

$$R_\odot = 8.0 \pm 0.5 \text{ kpc.} \tag{6.1}$$

The distance can be found, e.g., by measuring distances to globular clusters and determining the centroid of their distribution. (A more recent determination utilizes the observed

---

[1] An exception is a recently discovered population of intergalactic stars in galaxy clusters, comprising of order 10% of the stellar population in that environment. Also, the first stars formed in the Universe probably preceded the first galaxies.

Figure 6.1 Three examples of spiral galaxies, each with its disk at a different angle to our line of sight, from face-on (M51, top left), through inclined (NGC 3370, top right), to nearly edge-on (NGC 4594, bottom). Note the spiral pattern traced by star-forming regions and, in the edge-on case, the thin gas-and-dust disk that is discernible by its extinction of the starlight behind it. Note also the varying domination of the disk and bulge components among the three galaxies, and the spheroidal population of globular clusters, evident as numerous compact sources, in the edge-on case. Image scales are of order 20 kpc in each picture. Photo credits: S. Beckwith, A. Riess, NASA, ESA, and the Hubble Heritage Team.

angular sizes of orbits of stars of known velocity near the Galactic center). From the differential motion of stars in the solar neighborhood (as determined by proper motions and Doppler line-of-sight velocities), the orbital velocity of the Sun around the Galactic center is

$$v_\odot = 220 \text{ km s}^{-1}. \tag{6.2}$$

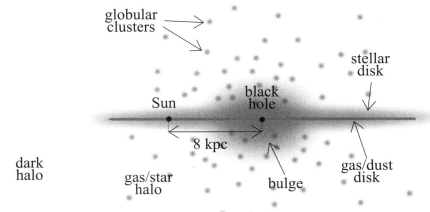

Figure 6.2 Schematic edge-on view of the main components of a spiral galaxy. The location of the Sun in the Milky Way is indicated.

The orbital period is therefore

$$\tau = \frac{2\pi R_\odot}{v_\odot} \sim 2 \times 10^8 \text{ yr}. \tag{6.3}$$

For a spherical mass distribution, the mass $M(R_\odot)$ internal to a radius $R_\odot$ obeys

$$\frac{GM(R_\odot)}{R_\odot^2} = \frac{v_\odot^2}{R_\odot}, \tag{6.4}$$

or

$$M(R_\odot) \approx \frac{R_\odot v_\odot^2}{G} = \frac{8 \text{ kpc} \times 3.1 \times 10^{21} \text{ cm kpc}^{-1} \times (2.2 \times 10^7 \text{ cm s}^{-1})^2}{6.7 \times 10^{-8} \text{ cgs}}$$

$$= 1.8 \times 10^{44} \text{ g} \approx 10^{11} M_\odot. \tag{6.5}$$

Although the mass distribution in the Milky Way is highly nonspherical, Eq. 6.4 is a good approximation also for a circular flattened mass distribution, in which $v(r) = $ constant, which is roughly true for the Milky Way. About half the mass interior to $R_\odot$ is in stars, and the typical stellar mass is about $0.5 M_\odot$, so there are about 10^{11} stars interior to the solar orbit.

6.1.1 Galactic Components

6.1.1.1 The Disk

The Galactic disk has a mass distribution that falls exponentially with both distance r from the center and height z above or below the plane of the disk:

Figure 6.3 *Top*: An optical-light image in the direction of the Milky Way's center. Although the Galactic disk is visible as a broad swath of light, the center of the Galaxy is obscured by the dust in the disk. *Bottom left:* A half-sky image (i.e., 2π steradians) with the Milky Way overhead. The disk + bulge structure can be seen. The region shown in the top panel is indicated. *Bottom right:* Optical image of the nearby edge-on spiral galaxy NGC 891. Note how similar the Milky Way and NGC 891 appear. Photo credits: Wei-Hao Wang; and C. Howk, B. Savage, N. A. Sharp, WIYN/NOAO/NSF, copyright WIYN Consortium, Inc., all rights reserved.

$$\rho(r, z) = \rho_0 \left[\exp\left(-\frac{r}{r_d}\right) \right] \left[\exp\left(-\frac{|z|}{h_d}\right) \right]. \tag{6.6}$$

The scale length of the disk, $r_d = 3.5 \pm 0.5$ kpc, and hence at $r = 8$ kpc the Sun is in the outer regions of the Galaxy. The characteristic scale height is $h_d = 330$ pc for the lower-mass (older) stars in the disk and $h_d = 160$ pc for the gas-and-dust disk. The Sun is located at $z = 30$ pc above the midplane of the disk. The mass of the disk within one scale radius is

$$M_{\mathrm{disk}} \sim 10^{10} M_\odot, \tag{6.7}$$

most of it in stars (and about 10% in gas). This implies about 2×10^{10} stars, and a mean density

$$\bar{n}_{\mathrm{stars}} \sim \frac{2 \times 10^{10}}{\pi (3500 \, \mathrm{pc})^2 \times 2 \times 330 \, \mathrm{pc}} \sim 1 \, \mathrm{pc}^{-3} \sim 3 \times 10^{-56} \, \mathrm{cm}^{-3}. \tag{6.8}$$

The mean distance between stars is then

$$\bar{d} = \bar{n}^{-1/3} \sim 1 \, \mathrm{pc} \tag{6.9}$$

Ignoring gravity for a moment, the mean free path for a star to physically collide with another star (i.e., for the centers of the stars to pass within two stellar radii of each other) is

$$l = \frac{1}{\bar{n}\sigma_{\mathrm{geom}}}, \tag{6.10}$$

where the geometric cross section is $\sigma_{\mathrm{geom}} = \pi (2r_\odot)^2$, for a star of solar radius. The random, noncircular, velocity of stars in the disk is typically $v_{\mathrm{ran}} \sim 20$ km s^{-1}. For a given star, the time between collisions is then

$$\tau_{\mathrm{coll}} = \frac{l}{v_{\mathrm{ran}}} = \frac{1}{\bar{n}\sigma_{\mathrm{geom}} v_{\mathrm{ran}}}$$

$$\sim \frac{1}{3 \times 10^{-56} \, \mathrm{cm}^{-3} \times \pi \times (2 \times 7 \times 10^{10} \, \mathrm{cm})^2 \times 2 \times 10^6 \, \mathrm{cm \, s}^{-1}}$$

$$= 2 \times 10^{26} \, \mathrm{s} = 7 \times 10^{18} \, \mathrm{yr}. \tag{6.11}$$

This is $\sim 5 \times 10^8$ times the age of the Universe, and thus, accounting only for the geometric cross section, most stars never collide (in a galaxy with 10^{11} stars, there will be of order 100 collisions in the course of 10^{10} yr). In reality, **gravitational focusing**—the fact that nearby stars attract each other—increases the effective cross section for a physical collision as

$$\sigma_{\mathrm{eff}} = \sigma_{\mathrm{geom}} \left(1 + \frac{v_e^2}{v_{\mathrm{ran}}^2}\right), \tag{6.12}$$

where v_e is the escape velocity from the surface of the star (see Problem 2 for a derivation). The escape velocity is defined as the velocity of a particle that has equal kinetic and potential energies, and is therefore not bound. For a Sun-like star, $v_e = 620$ km s^{-1}, and hence the effective collision cross section is ~1000 times larger than the geometrical one. This factor, however, does not change the fact that stellar collisions in the disk are extremely rare. In the innermost regions of the galaxy, where the density and random velocities are the highest, the probability for a stellar collision is significant. Also, if we consider the cross section to be that of a planetary system of radius 20 AU, the collision time goes down by a factor 1000^2 to about 10^{12} yr (gravitational focusing becomes unimportant in this case, because the escape velocity from 20 AU in only $v_e \sim 10$ km s^{-1}). Thus, about 1% of all disk stars undergo, during their lifetimes, a disruption of their planetary systems (if they have such systems) by the close passage of another star.

The gas-and-dust disk has a smaller scale height than the stellar disk. From our vantage point near the midplane of the Galactic disk, extinction by dust makes it extremely difficult to see most of the Galaxy at optical wavelengths (and to a lesser degree, also at IR wavelengths). The extinction to the Galactic center at visible wavelengths amounts to a factor of about 10^{11}. The view perpendicular to the disk, looking at the halo of the Galaxy and beyond our galaxy, is essentially unattenuated.

The spiral arms in the disk are the sites of slightly increased gas density. The conspicuous appearance of spiral arms is due to the fact that star formation is significantly enhanced along the arms, and the luminosity of young, massive stars is high. It is often remarked that O and B stars and H II regions trace out the spiral arms in the Milky Way and other spiral galaxies like "beads on a wire." The fact that there is ongoing star formation in the disk means that the stellar population in the disk has a large range of ages, including old stars (10 Gyr), intermediate age stars like the Sun (5 Gyr), and recently formed stars.

It is important to realize that the spiral arms are almost stationary features, through which stars move in and out as they orbit a galactic center, or transient features that appear and disappear at different locations in the disk. Recall the Sun's orbital period around the Galactic center, of 2×10^8 yr. Stars at half the galactocentric distance orbit the center with about the same velocity of 200 km s^{-1}, and so have a period that is half as long. If the spiral pattern were moving with the stars, the pattern would have wound about 50 times over 10^{10} yr, and would have been washed out. Instead, the spiral patterns are observed to wind just a few times, at most. One of the mechanisms proposed to explain spiral arms involves **density waves**, regions of higher stellar and gas density. A common example of a density wave is a traffic jam through which cars pass at lower speed, leading to a density enhancement along the road. The traffic jam can be stationary, or move backward or forward at some speed, which is unrelated to the speed of the cars going through it. A traffic jam can also disappear at some location along a road and appear at another. In a spiral galaxy disk, density waves can result from particular shapes of the orbits of the individual stars and the gas, which spend longer times during their orbits at the locations of the peaks of the density waves. Other mechanisms have also been suggested to produce spiral arms, such as interactions between galaxies. It is possible that several effects are at play.

6.1.1.2 The Spheroid

The spheroid of the Galaxy consists of a spheroidal stellar bulge of typical size 1 kpc, with a density proportional to radius as

$$\rho \sim r^{-3}, \tag{6.13}$$

and a stellar and gas halo with a similar profile out to about 50 kpc. About 200 globular clusters are also members of this halo population and they follow the same spatial distribution. The instrinsic three-dimensional shape of the spheroid, which at the arbitrary projection of any particular galaxy appears circular or elliptical, is probably that of a body of revolution obtained by rotating an ellipse about its minor axis. This is called an *oblate spheroid*.

The stars in the bulge and in the halo are old (10–14 Gyr). The age of the stellar population in a globular cluster, which is a system of stars that formed together at about the same time, can be deduced from its H-R diagram, by the location of the turnoff from the main sequence to the giant branch (see Fig. 4.1). Stellar models are used to predict at what age the spectral types at the observed turnoff point complete their main-sequence lifetimes. The spectra of the stellar atmospheres also reveal the element abundances that the stars had when they formed. The old stars comprising the halo population formed from material that underwent relatively little heavy-element enrichment by the winds from giants and the ejecta from supernovae, following previous stellar generations. For example, in halo stars the iron-to-hydrogen mass fraction, relative to solar, is

$$\frac{[\text{Fe/H}]}{[\text{Fe/H}]_\odot} = 10^{-4.5} \text{ to } 10^{-0.5}. \tag{6.14}$$

6.1.1.3 The Galactic Center

The Galactic center has been studied intensively in recent years at IR, radio, X-ray, and γ-ray wavelengths, which are less affected than optical light by the large amounts of dust on our line of sight. The Galactic center region has a large density of stars, and includes several young star clusters, supernova remnants, and a complex assortment of atomic and molecular gas (see Fig. 6.4). Radio and X-ray observations reveal a compact, stationary source named Sagittarius A* at the location of the kinematic center (i.e., the center as determined from stellar orbits). The high angular resolution possible with radio-telescope interferometers reveals a size of $\sim$1 AU for this object. Observations over the past decade of the actual orbits of stars around the center (see Fig. 6.5) have allowed a fairly accurate measurement of the mass of this object, $M \approx 4 \times 10^6 M_\odot$. The concentration of such a large mass within such a small region, with little luminosity to show for it, is strong evidence for the presence of a *supermassive black hole*. The radio and X-ray emission are the result (in a manner that is not yet fully understood) of the release of gravitational energy of matter being accreted by the black hole. Such objects appear to be present in the centers of all large galaxies, based on the observed motions of stars and gas in the central regions. The black-hole masses are roughly proportional to the stellar masses of the bulges, and have a range of about 10^6–10^9 $M_\odot$. It is currently not known whether central black holes are by-products of galactic formation and evolution, or perhaps play an

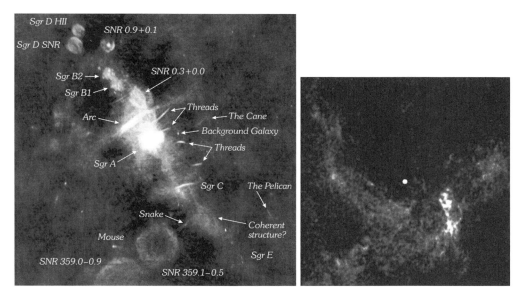

Figure 6.4 *Left*: A radio-wavelength (90 cm) image of ~1 kpc around the Galactic center. The Galactic plane goes from top-left to bottom-right (compare to Fig 6.3). The central source radio source, Sagittarius A, is marked. The unresolved radio source at the kinematic center of the Galaxy, Sagittarius A*, is within it. A number of other structures are also identified, mainly supernova remnants emitting synchrotron radiation and H II regions that emit free–free radiation. *Right*: Radio image at a wavelength of 1.3 cm of the central parsec around Sagittarius A*, which is at the center of the image, surrounded by diffuse emission from ionized gas. Photo credits: NRAO/AUI and N. E. Kassim, Naval Research Laboratory; NRAO/AUI, Jun-Hui Zhao, and W. M. Goss.

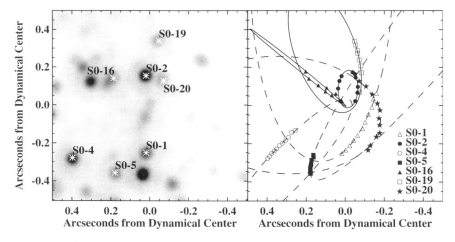

Figure 6.5 *Left:* Near-infrared (2.2-μm-wavelength) diffraction-limited image of the central arcsecond (0.04 pc) around the Galactic center, obtained in June 2005. *Right:* Positions of the seven stars marked in the left panel, as traced with such images between 1995 and 2005, and the best-fitting elliptical orbits. The orbits indicate an enclosed mass of $\approx 4 \times 10^6 M_\odot$ within a radius of <45 AU (the closest approach achieved by star S0-16), presumably a supermassive black hole at the position of Sagittarius A*. Figure credit: A. Ghez and J. R. Lu, see *Astrophys. J.* (2005), 620, 744.

important role in regulating galaxy growth. We will return to supermassive black holes in section 6.3.

6.1.1.4 Cosmic Rays

Cosmic Rays is a term for energetic matter particles that arrive at Earth from space. They can be detected directly using high-altitude balloon experiments, or by tracking the showers of secondary particles (mainly muons) that they produce when they hit Earth's atmosphere. Cosmic rays consist of electrons, protons, and heavier nuclei, with all of the elements present. Some cosmic rays are of solar origin, from particles carried out with the solar wind. Other cosmic ray particles are Galactic, and are thought to be produced and accelerated mainly in supernova explosions. They are then trapped by the magnetic field of the Milky Way's disk. Finally, some cosmic rays are probably of extragalactic origin, and come from active galactic nuclei (see section 6.3) and from gamma-ray bursts.

6.1.1.5 The Dark Halo

The last galactic component we will discuss, and in some detail, is the dark halo. The evidence for dark halos in galaxies is kinematic, based on the measurement of **rotation curves**, i.e., the circular velocity, $v(r)$, of test particles in orbit around the center of a galaxy, as a function of radius r. The circular velocity is found by observing material (e.g., gas) that is orbiting in the gravitational potential of the galaxy and measuring the line-of-sight velocity from the Doppler shift of a known transition, seen in the spectrum in emission or absorption. The measurement then needs to be corrected for various observational effects, such as inclination, motion of the observer, and projection (since the observer sees only the velocity component along the line of sight), to obtain the intrinsic circular velocity. Rotation curves are often measured using the 21-cm emission line of H I, the Balmer Hα emission line from H II regions, the absorption lines in the integrated light from stars, and the molecular emission lines from CO. Figure 6.6 shows a typical galactic rotation curve.

A remarkable finding that has few or no exceptions is that, at large enough radii, galaxy rotation curves are "flat," i.e., $v(r) = $ constant. This holds even in the very outer parts of galaxies, far beyond the regions where most of the emission from stars and from gas is concentrated, and where only trace amounts of luminous matter exist (some luminous matter is generally needed, in the role of test particles, to make the kinematic measurements). If the mass distribution followed the light, one would expect that, at some radius, most of a galaxy's mass would be within, and hence the rotation curve would fall in an approximately Keplerian fashion, $v \sim r^{-1/2}$, as for a point mass. Since

$$M(r) = \frac{v^2 r}{G},$$

(6.15)

$v(r) = $ constant implies

$$M(r) \sim r.$$

(6.16)

Thus, the flat rotation curves mean that, far beyond the exponentially dimming disk and the spheroid with its steeply falling light distribution, the mass enclosed within a radius r continues to grow linearly with r, out to the largest radii that can be measured. This dark halo constitutes much of the galaxy's total mass, the exact fraction depending on where

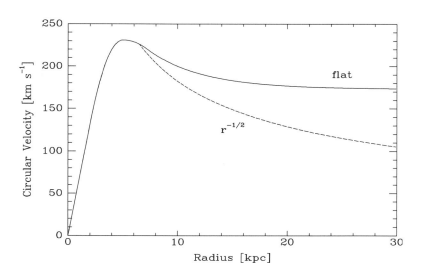

Figure 6.6 Schematic example of a spiral galaxy's rotation curve—its circular velocity as a function of radius. Outside the inner few kiloparsecs, rotation curves invariably become "flat," i.e., approach a constant velocity, suggesting the presence of a halo of dark matter. The dashed curve shows the $\sim r^{-1/2}$ Keplerian falloff that would be expected, in the absence of dark matter, beyond a radius that includes most of the visible luminous matter.

the dark halo ends. For example, in the Milky Way, the flat rotation curve is seen to extend out to $\sim$30 kpc, of order 10 disk scale lengths. Thus, the total mass is about 10 times the mass inside a scale length, and 80–90% of the Galaxy's mass, of order $10^{12} M_\odot$, is dark. The luminous parts of galaxies are just the "tip of the iceberg," while the majority of the mass is dark and extended. Combining Eq. 6.16 and the scaling relation between mass, density, and radius, we get

$$r \sim M(r) \sim \rho r^3, \tag{6.17}$$

and hence in the outer regions of a galaxy, where the dark matter is dominant, the mass density profile of the dark matter must be

$$\rho(r) \sim \frac{1}{r^2}. \tag{6.18}$$

6.1.2 The Nature of the Dark Matter

Let us consider the possible nature and forms of this ubiquitous dark matter.

Gas: If the dark matter were a gas of normal baryonic matter, it could be one of the following:

  atomic gas—but it would then radiate strongly at 21 cm.

  molecular gas—possible, in principle, but the gas would have to be un-enriched in heavy elements (to avoid emission from CO), and very cold (to avoid H_2 emission). It could still be detected by the effects of H_2 absorption and scattering of background UV sources, and some first attempts to see such an effect have not succeeded, making this option questionable. However, this dark matter candidate has not been fully ruled out yet.

ionized gas—a hot ionized gas is a strong emitter in X-rays of thermal free-free radiation[2], which is not observed.

Dust: Dust would emit IR radiation, and would produce visible extinction of background objects by the halos surrounding other galaxies. Furthermore, dust is composed of "metals" (i.e., elements heavier than helium), and on average only about 2% of baryonic matter is composed of metals. One would have to somehow drive away from the galaxy a mass of hydrogen and helium gas that is 50 times greater than the mass of the dust making up the halo.

Massive Compact Halo Objects (MACHOs): This category includes all gravitationally bound, star-like, objects, including planets and black holes. The various possibilities are as follows:

main-sequence stars—cannot comprise the dark halos of galaxies since they would be visible, not dark.

giant stars—are even more luminous than main sequence stars, so are certainly not an option.

neutron stars—their formation is accompanied by supernova explosions, which produce large amounts of heavy elements. If much of the dark halo were composed of neutron stars, the metal abundances in the halo gas would be very high, but they are measured to be low. Furthermore, only a fraction (about 1/10) of the initial mass of a massive star ends up in a neutron star, and the rest would have to be blown away from the galaxy.

black holes—the same argument as for neutron stars can be made. However, so little is known about the supernova explosions (if any) that accompany black hole formation that the argument does not carry so much weight. The distribution of binary separations observed in the halo can also be used to argue indirectly against a dark halo composed of black holes, since encounters with the black holes would disrupt the binaries.

white dwarfs—if dark halos are composed of white dwarfs, the winds from the giants that preceded their formation would have produced a large enrichment in intermediate-mass elements (He, N, Ne, C, O) of the halo gas, which is again not seen. Furthermore, one would expect to see luminous red halos, composed of the red-giant precursors of the white dwarfs, when looking at distant galaxies (corresponding to large *lookback times*, before the white dwarfs formed), and these are probably not seen. Due to the incomplete theoretical understanding of element enrichment by stellar winds, and the difficulty of observing very distant galaxies, this option is not completely ruled out.

brown dwarfs and "planets"—such objects, with masses in the range between the planet Jupiter ($M_J = 0.001 M_\odot$) and the stellar ignition limit ($M_{min} \approx 0.07 M_\odot = 70 M_J$), emit little radiation and cause no enrichment of the interstellar medium. They were therefore leading

---

[2] *Thermal bremsstrahlung* is emitted by the electrons in an ionized gas that has a Maxwell-Boltzmann velocity distribution, even if the gas is tenuous enough for the radiation to escape freely with few scatterings, and hence the radiation is not in thermal equilibrium with the gas.

contenders for MACHO-type dark matter, until they were ruled out by the gravitational lensing experiments we will describe below.

Elementary Particles: This category includes any type(s) of elementary particles, known or unknown. The following are some options:

electrons and protons—this is just ionized hydrogen, which we already excluded based on the low observed level of X-ray emission from galactic halos.

neutrons—free neutrons decay in about 15 min into electrons and protons (plus neutrinos), which are excluded.

massive neutrinos—although it now seems likely that neutrinos have mass, it is not clear if they are massive enough to be a significant component of the dark matter. Their relativistic speeds also pose a problem in models of galaxy formation if the neutrinos constitute the main dark-mass component. However, this option has not been fully excluded.

cold dark matter—this subcategory refers to a broad class of currently undiscovered, but theoretically proposed, particles. These particles come from extensions to the "standard model" of particle physics but are not present in the standard model itself. They all share the property that they are "weakly interacting" (like neutrinos), i.e., they are affected only by the weak and gravitational interactions, and hence do not emit electromagnetic radiation. Furthermore, they have a high mass per particle, and since they are bound in a galaxy potential, they move at nonrelativistic speeds and are thus "cold" (as opposed to the "hot" neutrinos). Some candidate particles are "axions" and particles that are predicted to exist in the framework of a theory called supersymmetry. The whole zoo of plausible particle dark matter candidates is sometimes called WIMPS, weakly interacting massive particles.

Reviewing the above list of dark matter candidates, two categories are neither excluded nor arguably unlikely: cold dark matter and sub-stellar-mass MACHOs. The MACHO option has been recently tested experimentally via the phenomenon of **gravitational lensing**. To understand how this was done, let us digress and develop the theory of gravitational lenses.

6.1.3 Gravitational Lensing Basics

General relativity predicts that space is curved by the presence of mass. As we have seen in the example of the Schwarzschild metric, in a gravitational field, as seen by a distant observer, light can be thought of as moving at a speed less than c, and the field thus acts effectively as an index of refraction. Therefore, a light ray traversing a region where the gravitational field has a gradient (e.g., near a point mass) will **bend** toward the mass, in accordance with the Fermat principle (of which Snell's law in optics is an example).

Consider, then, a light ray from a distant source approaching a point mass M and being bent as a result of this effect (see Fig. 6.7). Denote the distance of closest approach of the ray to the mass (the *impact parameter*) as b, and the bending angle of the ray as α. Let us assume that the following two conditions apply: the **gravitational field is weak**, even at the

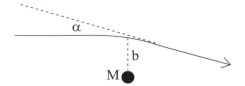

Figure 6.7 Gravitational deflection of a light ray from a distant source passing a point mass M to within a distance b, and continuing in the direction of a distant observer. In the weak-field limit (i.e., if, at the closest point of approach the gravitational potential $\phi \ll c^2$, or equivalently, $b \gg r_s$), the deflection angle is $\alpha = 4GM/c^2b$.

location of the ray's closest approach to the mass—this means that b is much larger than the Schwarzschild radius, or

$$\frac{GM}{c^2b} \ll 1; \tag{6.19}$$

and the radius over which the mass has significant bending influence on the the ray, namely $\sim b$, is much less than the other relevant distances in the problem, namely the distances between the observer, the source, and the point mass. This is called the **thin-lens approximation**, again in analogy to optics. Similarly, we shall refer to the point mass as the "lens" due to the same analogy. A calculation (which we will not pursue) of a photon trajectory in the Schwarzschild metric under these limiting conditions gives a simple formula for the bending angle,

$$\alpha = \frac{4GM}{c^2b} = \frac{2r_s}{b}. \tag{6.20}$$

Note that α is proportional to M and inversely proportional to b.

For example, for a light ray coming from a distant star projected near the limb of the Sun,

$$b \approx r_\odot = 7 \times 10^{10} \text{ cm}. \tag{6.21}$$

The weak-field and thin-lens approximations certainly apply ($b \approx r_\odot \sim 10^5 r_s$ and $d_\odot \sim 200b$), and therefore Eq. 6.20 gives

$$\alpha = \frac{2r_s}{b} = \frac{2 \times 3 \times 10^5 \text{cm}}{7 \times 10^{10}\text{cm}} = \frac{6}{7}10^{-5} \text{ rad} = 1.8 \text{ arcsec}, \tag{6.22}$$

where we have converted radians into arcseconds, by multiplying by $3600 \times 180/\pi$ (recall that an arcsecond is 1/3600 of a degree). Thus, light rays from stars projected near the solar limb will bend by this angle, and will appear to an observer on Earth to be arriving from a direction further from the limb. In other words, when the Sun is near the line of sight to these stars their positions on the sky will appear to shift away from the limb, relative to their positions when the Sun is elsewhere.[3] This phenomenon is an example of **lensing** of stars by the Sun's gravitational potential.

The first attempts to measure the lensing effect were carried out in 1919, during a total solar eclipse, when the Moon covers the Sun and thus reduces its glare, allowing stars to be seen projected near the limb in daytime. These were the first experimental tests of general relativity. Although the results were not conclusive due to the difficulty of measuring such a small angular shift under transient and difficult conditions, the results were widely publicized by the popular press, and turned Einstein overnight into a celebrity. Since then, this and many other predictions of the weak-field limit of general relativity have been confirmed experimentally to great accuracy.

To proceed, consider now the following three elements: a point light source, a point mass (i.e., a "lens"), and an observer, with all three on a straight line (see Fig. 6.8). In optical terms, the source is on the "optical axis" of the lens. Clearly, only rays with a particular impact parameter b will reach the observer—rays with larger b will not be deflected enough, and rays with smaller b will be deflected too much. Thus, all rays emitted at a certain angle to the optical axis will reach the observer, and by symmetry, the source will appear deformed into a ring, called an **Einstein ring**, with an angular radius θ_E, and physical radius R_E in the plane of the lens. Let us label the observer–source distance D_{os}, the observer–lens distance D_{ol}, and the lens–source distance D_{ls}. Since we assumed the weak-field approximation, $GM/c^2 b \ll 1$, then $\alpha = 4GM/c^2 b \ll 1$, i.e., α is a small angle. Similarly, the thin-lens approximation, $b \ll D_{ol}$, implies $\theta_E \ll 1$. Looking at the line segment SI, we can therefore write

$$SI = \alpha D_{ls} = \theta_E D_{os}. \tag{6.23}$$

Substituting Eq. 6.20 for α, with $b = R_E = D_{ol}\theta_E$, gives

$$\frac{4GM}{c^2 D_{ol}\theta_E} D_{ls} = \theta_E D_{os}, \tag{6.24}$$

or

$$\theta_E = \left(\frac{4GM}{c^2} \frac{D_{ls}}{D_{ol} D_{os}} \right)^{1/2}. \tag{6.25}$$

The **Einstein angle** sets the angular scale of the problem. For example, if we take a lens approximately halfway between the observer and source, then $D_{ls}/D_{os} \approx 1/2$, and

$$\theta_E = 0.64 \times 10^{-3} \text{ arcsec} \left(\frac{M}{M_\odot} \right)^{1/2} \left(\frac{D_{ol}}{10 \text{ kpc}} \right)^{-1/2}. \tag{6.26}$$

---

[3] Atmospheric refraction produces a similar effect on the setting Sun, giving it a flattened shape, and making it visible above the horizon when in reality it has already set. Note, however, that contrary to atmospheric refraction, gravitational light deflection does not depend on the wavelength of the photons, i.e., it is *achromatic*; the equivalence principle of general relativity, which states that a reference frame in a gravitational field cannot be distinguished from an accelerated reference frame, dictates that photons of all energies will be deflected by the same angle given by Eq. 6.20.

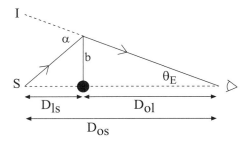

Figure 6.8 Geometrical configuration in which a point source of light (S), a lensing point mass, and an observer (on the right-hand side) are on a straight line, with distances as marked. For a particular lensing mass, M, only rays passing by the lens at a specific impact parameter b will reach the observer. By symmetry, the image of the source will be deformed into a ring, of angular radius θ_E (see Eq. 6.27), called an Einstein ring. The large angles drawn are only for the sake of clarity, and in practice all angles are extremely small. Light rays are shown to bend at one point along their path, in keeping with the "thin-lens" approximation.

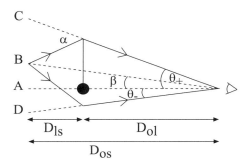

Figure 6.9 Geometry in which the point source of light (B), is off the "optical axis" (A) defined by the lens and the observer. In this case, two lensed images are seen by the observer. One image is at position C in the source plane, at an angle θ_+ from the lens. The second lensed image is on the opposite side of the lens, at position D in the source plane, separated from the lens by an angle θ_-.

Thus, a stellar-mass object halfway on our line of sight to a star at a distance of 20 kpc will gravitationally lens and deform the light from the star into an approximately 1-milliarcsecond Einstein ring. If the lens is a stellar-mass object in a galaxy at a "cosmological" distance of 1 Gpc (10^9 pc), then the Einstein angle becomes of order a microarcsecond. This is the reason why lensing by stellar-mass objects is often called **microlensing** (a name that is retained also when the lens is closer, and the angular scale is actually larger).

Consider now a more general case, where the light source is not on the optical axis of the lens, but rather at some angle β to it (Fig. 6.9). As before, $\alpha \ll 1$, and the thin-lens approximation, $b \ll D_{ol}$, implies $\theta \ll 1$. Since $\beta < \theta$, β is also a small angle. From Fig. 6.9, we can see that $AB + BC = AC$. Considering the small angles projected on these line segments, we can write

$$D_{ls}\alpha + D_{os}\beta = D_{os}\theta. \tag{6.27}$$

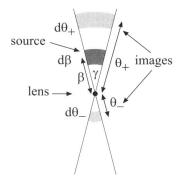

Figure 6.10 Schematic illustration showing the actual location on the sky of an unlensed source, at an angle β from the lens, and its two apparent lensed images, at positions θ_+ and θ_-, demonstrating the calculation of the magnification of the lensed images. The source has an extent $d\beta$ in the radial angular direction and $\beta\gamma$ in the tangential direction. The lensed image at position θ_+ is tangentially stretched by the ratio θ_+/β. The lensed image can also be either stretched or shrunk in the radial angular direction, by the ratio $d\theta_+/d\beta$. The magnifications, i.e., the ratios of solid angles subtended by the images and by the source, are therefore $\theta_\pm d\theta_\pm/\beta d\beta$. Since surface brightness is conserved, the magnification of each image is also its amplification.

Isolating β, we find

$$\beta = \theta - \frac{D_{ls}}{D_{os}}\alpha = \theta - \frac{D_{ls}}{D_{os}}\frac{4GM}{c^2 D_{ol}\theta} = \theta - \frac{\theta_E^2}{\theta}, \tag{6.28}$$

or,

$$\theta^2 - \beta\theta - \theta_E^2 = 0. \tag{6.29}$$

Solving this quadratic equation for θ, the angle of the **lensed image** relative to the lens, we find two solutions,

$$\theta_\pm = \tfrac{1}{2}[\beta \pm (\beta^2 + 4\theta_E^2)^{1/2}]. \tag{6.30}$$

Thus, there will always be **two lensed images** for a point-mass lens (except when the source is on the optical axis, $\beta = 0$, in which case we recover the single-image, Einstein-ring solution). The two images have θ with opposite signs, and hence they will straddle the lens on both sides.

Next, let us calculate the appearance of the lensed source on the sky. The source, were it not lensed, would subtend a tangential angle γ relative to the origin at the position of the lens, and a radial angle $d\beta$ (see Fig. 6.10). The lensing effect shifts the images of the source on the sky radially, relative to the lens, and therefore produces tangential stretching of the images, by factors $\theta_\pm/\beta$. These ratios can be calculated directly from Eq. 6.30. At the same time, the radial angular width of the source can be stretched or shrunk in the lensed images relative to the original size, by a factor $d\theta_\pm/d\beta$. These derivatives are, again, easily obtained from Eq. 6.30. The relative increase in the angular size, or **magnification**, of each image will therefore be

$$a_\pm = \frac{\theta_\pm d\theta_\pm}{\beta d\beta} = \frac{\theta_\pm}{2\beta} \left[1 \pm \frac{\beta}{(\beta^2 + 4\theta_E^2)^{1/2}} \right]. \tag{6.31}$$

Surface brightness (i.e., flux per solid angle of sky) is conserved by gravitational lensing, and therefore the magnification implies **amplification**, i.e., a magnified image will appear brighter. Microlensing occurs on angular scales such that the individual images cannot be resolved (i.e., discerned individually) by optical telescopes. However, one can measure the total magnification (or amplification), relative to the unlensed solid angle (or flux), as a result of the lensing:

$$a_{\text{tot}} = a_+ + a_- = \frac{u^2 + 2}{u(u^2 + 4)^{1/2}}, \tag{6.32}$$

where

$$u \equiv \frac{\beta}{\theta_E}, \tag{6.33}$$

i.e., the angular separation between source and lens in units of the Einstein angle. When u is large (i.e., the source is far, in projection, from the lens), we can see that a_{tot} tends to 1, as expected. When $u = 1$ (i.e., the source is at a position on the sky through which the Einstein ring would run, if the source were on the optical axis), then

$$a_{\text{tot}}(u = 1) = \frac{3}{\sqrt{5}} = 1.34, \tag{6.34}$$

i.e., the source will be magnified (amplified) by 34%. We can also see from Eq. 6.32 that for small u,

$$a_{\text{tot}} \sim \frac{1}{u}, \tag{6.35}$$

and formally becomes infinite when $u = 0$. This simply reflects the fact that a point source is being stretched into a circle, and there are an infinite number of points on a circle. In reality, there are no true "point sources"; every source of light has a finite physical (and hence angular) extent, and only one point in the source, which has zero area and hence emits zero flux, will be exactly at $u = 0$. The magnified flux in the observed Einstein ring will therefore be finite.

Now, let us add to the problem a relative transverse motion between the source and the lens, as viewed by an observer (such motion can be due mainly to the motion of the observer), with a velocity v measured in the plane of the lens. The simplest case is apparent motion along a straight line on the sky. From Fig. 6.11, we see that the source angle will now be a function of time,

$$\beta(t) = \left[\beta_0^2 + \frac{v^2}{D_{\text{ol}}^2}(t - t_0)^2 \right]^{1/2}, \tag{6.36}$$

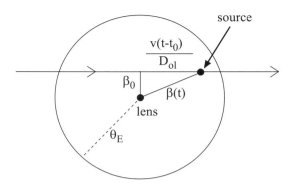

Figure 6.11 Relative motion on the sky of a lens (center, with its Einstein angle θ_E indicated) and a source that is projected behind it. Projected to the plane of the lens, the source moves at a constant transverse velocity v in a straight line, with an angle of closest approach β_0 at time t_0. From the Pythagorean theorem, the time dependence of the lens–source separation, $\beta(t)$, is given by Eq. 6.36.

where β_0 is the angle of closest approach and t_0 is the time of closest approach, or

$$u(t) \equiv \frac{\beta(t)}{\theta_E} = \left[u_0^2 + \frac{(t - t_0)^2}{\tau^2} \right]^{1/2}, \tag{6.37}$$

where

$$u_0 \equiv \frac{\beta_0}{\theta_E}, \quad \tau \equiv \frac{\theta_E D_{\text{ol}}}{v}. \tag{6.38}$$

Thus, the total magnification as a function of time, $a_{\text{tot}}(t)$, is obtained by putting $u(t)$ from Eq. 6.37 in place of u in Eq. 6.32. Since u is an even function of $(t - t_0)$, $a_{\text{tot}}(t)$ describes a curve that is symmetric in time with respect to t_0, starting and ending at $a_{\text{tot}}(t) = 1$ for $t = \pm\infty$, and rising on an "Einstein-ring crossing timescale" τ to the maximum at $t = t_0$ (see Fig. 6.12).

If such a microlensing magnification curve is observed in a source undergoing transient lensing, one can deduce the parameters describing the curve: u_0, t_0, and τ. The parameter τ is a function of v, θ_E, and D_{ol}, and therefore of v, M, and D_{ol} (assuming the source distance, D_{os}, is known). Thus, given v and D_{ol}, one can deduce the mass of the lens by measuring the characteristic time, τ, of the total magnification curve.

6.1.4 The Magellanic Cloud Microlensing Experiments

The discussion above shows that it is possible, in principle, to detect the presence of dark moving compact masses, and to deduce their properties, by virtue of their transient gravitational magnification effect on background sources. This led Paczyński in 1986 to propose an elegant experiment, which would test if the dark halo of the Milky Way is composed of MACHOs, as follows. Consider the Large Magellanic Cloud (LMC), which is a small satellite galaxy of the Milky Way, at a distance of $l = 50$ kpc. A virtue of the LMC

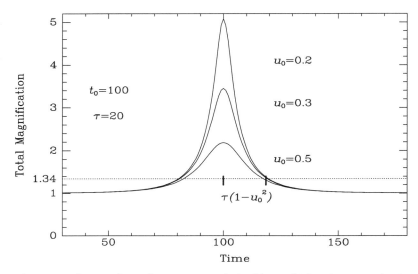

Figure 6.12 Theoretical microlensing curves, obtained by combining Eqs. 6.32 and 6.37, and showing the total magnification of a microlensing event as a function of time. Curves are shown for three different Einstein-angle-normalized impact angles u_0. The time of maximal approach is $t_0 = 100$ and the Einstein-crossing timescale is $\tau = 20$ (both in arbitrary time units). Note how the peak magnification rises with decreasing u_0, and equals approximately u_0^{-1}, as expected from Eq. 6.35.

is that it is near enough that many of its constituent stars can be discerned individually in telescope images (see Fig. 6.13). Assume that the Milky Way's dark halo, with a total mass M_{DM}, extends out to the LMC and is composed of MACHOs, each of mass m. The mean MACHO number density is then

$$n = \frac{M_{\mathrm{DM}}}{m\frac{4}{3}\pi l^3}.\tag{6.39}$$

At a given moment, a star in the LMC will be amplified by >1.34 if it happens to be projected within the Einstein angle θ_E of an intervening MACHO. We can therefore define a *lensing cross section*, for amplification by a factor of 1.34 or more, of

$$\sigma = \pi\left(\theta_E D_{\mathrm{ol}}\right)^2 = \pi\frac{4Gm}{c^2}\frac{D_{\mathrm{ls}}}{lD_{\mathrm{ol}}}D_{\mathrm{ol}}^2.\tag{6.40}$$

The number of such "targets" that will be crossed by any line of sight to a star in the LMC is

$$N \approx n\sigma l.\tag{6.41}$$

(In reality, both n and σ change with distance, so the accurate expression is

$$N = \int_0^l n(r)\sigma(r)dr,\tag{6.42}$$

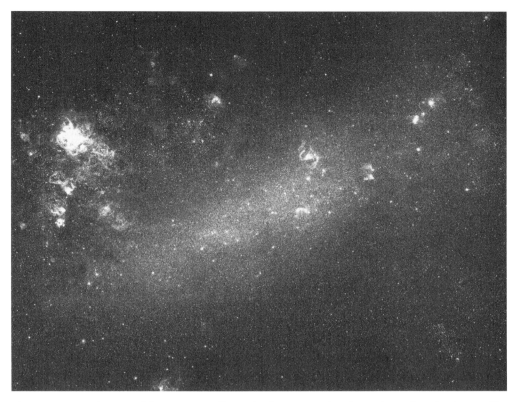

Figure 6.13 Optical image of the Large Magellanic Cloud (LMC), a "dwarf irregular" galaxy that is a satellite galaxy of the Milky Way. At a distance of 50 kpc, the LMC is close enough that the individual stars in it can be resolved from each other in ground-based telescope images, despite the limited resolution due to blurring in the Earth's atmosphere. Image width is about 5 kpc. Photo credit: NOAO/AURA/NSF.

but the approximate expression is sufficient for our purposes.) Inserting the expressions for n and σ into Eq. 6.41 gives

$$N \approx \frac{M_{\mathrm{DM}}}{m \frac{4}{3}\pi l^3} \pi \frac{4Gm}{c^2} \frac{D_{\mathrm{ls}}}{l D_{\mathrm{ol}}} D_{\mathrm{ol}}^2 l. \tag{6.43}$$

After canceling and taking the typical MACHO to be about halfway between us and the LMC, so that $D_{\mathrm{ls}} \sim D_{\mathrm{ol}} \sim l/2$, we find

$$N \approx \frac{GM_{\mathrm{DM}}}{lc^2} = \left(\frac{v_\odot}{c}\right)^2. \tag{6.44}$$

In the last equality we have used the fact that $GM(r)/r$ equals $[v(r)]^2$, and, since the rotation curve is flat, $v(r)$ is just $v_\odot$. We thus get the elegant result that, if the dark halo of the Galaxy (which constitutes most of the Galaxy's mass) is composed of MACHOs, the number of MACHOs along the sightline to an outside light source, and that are passed to within an Einstein angle, is just $(v_\odot/c)^2$. Since this number is obviously much smaller than one, it is actually the fraction of stars that are lensed, or the probability that a particular star is lensed. Note that N depends on the total mass M_{DM} (or alternatively, on the circular velocity $v_\odot$), but not on the individual MACHO masses m. Taking $v_\odot = 220$ km s^{-1}, we find

$$N \sim \left(\frac{220 \times 10^5 \text{ cm s}^{-1}}{3 \times 10^{10} \text{ cm s}^{-1}}\right)^2 \sim 10^{-6}. \tag{6.45}$$

In other words, at any given moment, one in a million stars in the LMC should be significantly lensed (i.e., amplified by more than 1.34) by a MACHO in our dark halo.

The Einstein angle of a solar-mass MACHO located at half the distance to the LMC (Eq. 6.26) is $\theta_E \sim 2 \times 10^{-9}$, and the apparent motion on the sky between the MACHO and the LMC star is due mainly to the Sun's orbit around the Galaxy center, $v = 220$ km s^{-1}. The Einstein *diameter* crossing timescale of a **lensing event** due to such a MACHO is then

$$\tau(1M_\odot) = \frac{2\theta_E D_{ol}}{v} = \frac{4 \times 10^{-9} \times 25 \times 3.1 \times 10^{21} \text{ cm}}{220 \times 10^5 \text{ cm s}^{-1}}$$
$$= 1.4 \times 10^7 \text{ s} \approx 6 \text{ months.} \tag{6.46}$$

Alternatively, if the halo is composed of Jupiter-mass MACHOs of mass $10^{-3} M_\odot$, then θ_E, and hence τ will be about 30 times smaller:

$$\tau(1M_J) = 6 \text{ days.} \tag{6.47}$$

The proposed experiment is then the following. Monitor regularly the brightnesses of about 10^7 LMC stars for about 5 years. If the Milky Way dark halo is made of MACHOs, regardless of their masses, at any given moment about 10 of the monitored stars will be undergoing a lensing event. If, for example, the MACHOs are of about solar mass, about 10×5 yrs $\times 2$ yr$^{-1} = 100$ lensing events with 6-month timescales will be discovered in the course of the 5-year survey. If the MACHOs are Jupiters, about 10×5 yr $\times 52$ yr$^{-1} = 2600$ week-long lensing events will be found. If the dark halo is not made of MACHOs, few lensing events will be found.

In 1991, several observational groups began carrying out just such experiments on dedicated $\sim$1-m-diameter telescopes. The experiments had just become feasible thanks to two technological developments. Large-area electronic CCD cameras allowed recording simultaneously the images of many LMC stars in a single frame. Computing power made it possible to automatically measure the light from millions of individual stars in these images and to search for the handful undergoing amplification by lensing, with the time dependence given by Eqs. 6.32 and 6.37. After the first year, two of the experiments reported the discovery of the first lensing events, having the expected time dependence. The Einstein-crossing timescales were $\tau = 35$–230 days, corresponding to MACHO masses of $m \sim (0.1$–$1)M_\odot$, assuming typical Galactic velocities and lenses about halfway along the line of sight to the sources. However, after 6 years of monitoring about 10 million stars, only 15, or so, lensing events had been found by each experiment, as opposed to the $\sim$100 that would have been expected, given these timescales.

MACHOs can therefore constitute no more than a small fraction of the Milky Way dark halo. A more thorough calculation, accounting also for the detection efficiencies of each

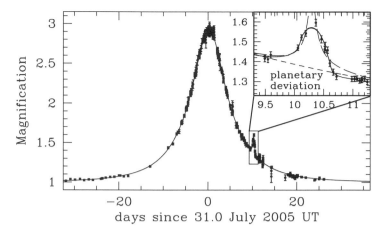

Figure 6.14 Example of a microlensing event observed in the direction of the Galactic bulge, involving a foreground star that temporarily magnifies the light from a background star projected behind it. Note the excellent fit (solid curve) to the data provided by Eqs. 6.32 and 6.37 during most of the event. However, the small, day-long, perturbation in the curve of magnification vs. time on its falling side (enlarged in the inset) reveals the presence of a planet around the lens star. A detailed model (solid curve in the inset) indicates an approximately 5-Earth-mass planet at an orbital radius of several AU from the main lens star. The single-lens model (short-dashed curve) and a model assuming a binary source star (long-dashed curve) are both ruled out by the data. Figure credit: PLANET, OGLE, and MOA collaborations, see J.-P. Beaulieu et al. 2006, *Nature*, 439, 437.

survey, indicated that about 20%, or less, of the dark halo is composed of MACHOs. If true, a 20% fraction in compact, about-solar-mass, objects poses a new and additional problem. It is much more than expected from the known stellar halo population of the Galaxy, and putting even this fraction in any of the known types of objects that have such masses (main-sequence stars, white dwarfs, neutron stars, black holes) raises similar problems to those already pointed out for the case that the entire halo were made of such objects. However, it is also possible, within the experimental and theoretical uncertainties, that the small number of observed microlensing events were all due to lensing by the known stellar populations in the Milky Way's disk and bulge and by stars in the LMC itself. The bulk of the dark matter must be made of something other than MACHOs, with cold dark matter now being the prime contender.

We note in passing that, although few microlensing events in the direction of the LMC have been seen, thousands of events toward the Galactic bulge have been recorded, due to lensing of a star by another star along the line of sight. Because of the large density of stars in this direction, the probability of star–star lensing is relatively high (see Problem 5). Figure 6.14 shows an example of data for such a microlensing event. A major motivation for discovering and closely following these events is a search for **extrasolar planets** around the lensing stars. Under some conditions, such a planet, even one with a small mass, can reveal its presence by means of the perturbation it causes in the magnification curve of the event. To date, a number of extrasolar planets have been discovered via microlensing, one of them in the event shown in Fig. 6.14.

6.1.5 Modified Physics Instead of Dark Matter

A perhaps more radical solution of the dark matter problem is that there is no dark matter, and it is the physics (which we are using to deduce kinematically that dark matter exists) that needs to be modified. Several such modifications to Newtonian mechanics have been proposed. Most notable among them has been Milgrom's and Bekenstein's *modified Newtonian dynamics* (MoND), which proposes that Newton's second law, $F = ma$, may not be applicable in the regime of the weak accelerations existing in galaxy halos. Newton's laws have been tested only in the Solar System, where, e.g., the Earth's centripetal acceleration is

$$a_\oplus = \frac{v_\oplus^2}{d_\odot} = \frac{(30 \times 10^5 \text{ cm s}^{-1})^2}{1.5 \times 10^{13} \text{ cm}} = 0.6 \text{ cm s}^{-2}. \tag{6.48}$$

By comparison, the acceleration of the Sun around the Galactic center is

$$a_\odot = \frac{v_\odot^2}{R_\odot} = \frac{(220 \times 10^5 \text{ cm s}^{-1})^2}{8 \times 3.1 \times 10^{21} \text{ cm}} = 2 \times 10^{-8} \text{ cm s}^{-2}, \tag{6.49}$$

more than seven orders of magnitude smaller than Solar System accelerations. In MoND, $F = ma$ is replaced by a modified acceleration law. The modified form includes a characteristic acceleration, $a_0 \approx 1 \times 10^{-8} \text{ cm s}^{-2}$, which is determined empirically by observations of galaxy rotation curves. At accelerations much smaller than a_0, the MoND version of the second law approaches the form

$$F = \frac{ma^2}{a_0}, \tag{6.50}$$

which leads to flat rotation curves (see Problem 7). MoND succeeds remarkably well in phenomenologically reproducing the rotation curves of galaxies. Relativistic versions of the theory (required, e.g., for making predictions of gravitational lensing phenomena) have been devised, but their observational implications are still being worked out. It is also debated whether or not the theory works well in all observed astronomical environments.

6.2 Galaxy Demographics

Three types of galaxies are observed to be common in the Universe. The dominant mass components of **spiral galaxies**, which we have discussed at length, are a bulge, a disk, and a dark halo. Different spiral galaxies have varying degrees of prominence of their bulges and disks, from those with a dominant bulge (often called "early-type" spiral galaxies) to

Figure 6.15 Several examples of elliptical galaxies (M49, M60, and M86), all at distances of ~15 Mpc. Photo credits: NOAO/AURA/NSF.

those with little or no bulge component (called "late-type" spirals).[4] Many spiral galaxies, perhaps including the Milky Way, also have a central stellar **bar**.

Elliptical galaxies resemble in many ways the bulges of spiral galaxies, but are devoid of a disk (see Fig. 6.15). The stars in elliptical galaxies are therefore predominantly old, as is the case in spiral bulges. As opposed to the stars in the disks of spiral galaxies, which move in approximately circular orbits in a well-defined plane around a galaxy's center, the stars in ellipticals move in "random" orbits, having a large range of inclinations and eccentricities.[5] The intrinsic shape of elliptical galaxies is not clear yet, and could be either a spheroid (i.e., the body of revolution obtained by rotating an ellipse around one of its axes) or a *triaxial ellipsoid*, of the form $x^2/a^2 + y^2/b^2 + z^2/c^2 = 1$. Being devoid of disks, ellipticals also have a low gas and dust content. The most massive known galaxies are ellipticals.

The most common galaxies are **irregular galaxies**, which constitute the third main type. The prototypical irregular galaxy is the LMC. Irregulars have ongoing star formation, and therefore young stellar populations. Both ellipticals and irregulars do not have well-formed disks, and it is therefore more difficult to determine whether they possess dark halos. It is not yet clear why these three main galaxy types exist, or how galaxy properties are established or evolve during galaxy formation, but various possible scenarios have been proposed.

Galaxies of all types come in a range of luminosities, masses, and sizes. Among these parameters, the total stellar luminosity of a galaxy can be measured most directly. The distribution of galaxy luminosities, or **luminosity function**, is described approximately by a *Schechter function*, giving the number, $\phi(L)$, of galaxies per unit volume with luminosity in an interval dL about L:

$$\phi(L)dL \approx \phi(L_*) \left(\frac{L}{L_*}\right)^{-1} \exp\left(-\frac{L}{L_*}\right) dL. \tag{6.51}$$

---

[4] This nomenclature has a historical, rather than a physical, origin, and is quite unfortunate since "early-type" galaxies consist of mostly "late-type" stars (i.e., low-mass stars), and "early-type" stars (i.e., massive stars) are found preferentially is "late-type" galaxies that have disks with ongoing star formation.

[5] It is important to point out here that closed Keplerian orbits are a rule only around spherically symmetric mass distributions. Galaxy potentials are, in general, not spherically symmetric, and therefore stars follow unclosed orbits that trace out a "rosette" that never repeats itself exactly.

Thus, there are fewer galaxies as one goes to higher luminosities, up to some characteristic luminosity, $L_* \approx 2 \times 10^{10} L_\odot$. The exponential cutoff above L_* means that galaxies with more than a few times L_* are very rare.

The Milky Way, as well as the nearest large galaxy, M31 in Andromeda, both have luminosities of roughly L_*. The mean density of galaxies with luminosities in the range $L_* \pm L_*/2$ is about

$$\phi(L_*) \sim 10^{-2} \text{ Mpc}^{-3}. \tag{6.52}$$

Thus, there is, on average, about one $\sim L_*$ galaxy per 100 Mpc^3, or an average distance of about 5 Mpc between large galaxies. We can use Eq. 6.11, which we derived to estimate the time between collisions of stars in a galaxy, to find roughly a typical time for physical collisions between galaxies. Taking a galactic radius of 50 kpc, a relative velocity between nearby galaxies of 500 km s^{-1}, and $\phi(L_*)$ for the density of galaxies,

$$\begin{aligned}
\tau_{\text{coll}} &= \frac{1}{n\sigma v} \\
&= \frac{1}{10^{-2} \times (3.1 \times 10^{24} \text{ cm})^{-3} \times \pi (2 \times 50 \times 3.1 \times 10^{21} \text{ cm})^2 \times 500 \times 10^5 \text{ cm s}^{-1}} \\
&\approx 5 \times 10^{12} \text{ yr}.
\end{aligned} \tag{6.53}$$

Thus, over the age of the Universe (about 10^{10} yr, which is also a typical age for most galaxies), about one in 500 large galaxies would undergo a collision with another large galaxy. In reality, the fraction is higher because galaxies are clustered in space (see below), and because of gravitational focusing. Furthermore, as we will see in chapter 8, in the past the density of galaxies, n, was higher than today. Several examples of such collisions are shown in Fig. 6.16. During a galaxy–galaxy collision, the constituent stars never physically collide with each other, for the same reason that the stars inside a given galaxy rarely collide—the cross sections are too small and the densities too low. Thus, in effect, colliding galaxies can "pass" through each other.

Note that, if a galaxy is in free fall toward a second galaxy, so are all of its constituent stars, gas, and dark matter, and therefore these components would not "feel" the pull of the second galaxy, were they all at similar distances from it. (For the same reason, a person on Earth does not feel the mean attraction of the Sun or the Moon, and an astronaut in orbit is weightless). However, the stars and the gas in one galaxy will feel the spatially varying gravitational field of the other galaxy, and will therefore experience tidal forces that can disrupt the stellar orbits, or tear away the stars completely from their parent galaxy. Examples of such tidal streams of stars and gas are seen in many colliding galaxies. This tidal work can take up enough of the kinetic energy of the relative motion between the galaxies that a collision can sometimes end as a **merger** of two galaxies. In fact, one idea for explaining the existence of different galaxy types is that ellipticals have resulted from the mergers of pairs of spirals. Furthermore, galaxy collisions apparently set off the collapse of gas clouds inside the galaxies, leading to vigorous bursts of star formation that are often observed in such interacting systems.

Figure 6.16 Two examples of pairs of galaxies in the process of collision, with "tails" of stars and gas that are drawn out of each galaxy by the tidal forces exerted by the other galaxy. *Top left:* The galaxy pair NGC 4038/9 in a montage of an optical image, tracing the stars, and a radio image in 21 cm, showing emission from neutral hydrogen. *Top right:* Zoom on the central region of NGC 4038/9, where the optical light output is dominated by H II regions and massive young stars formed as a result of the collision. *Bottom:* The colliding galaxy pair NGC 4676. Photo credits: NRAO/AUI and J. Hibbard; the European Southern Observatory; and NASA, H. Ford, G. Illingworth, M. Clampin, G. Hartig, the ACS Science Team, and ESA.

6.3 Active Galactic Nuclei and Quasars

We have seen that massive black holes, of 10^6–$10^9 M_\odot$, are common in the centers of large galaxies. There is no shortage of gas in these regions—abundant material is shed by normal stars during their evolution, and likely also by stars that have passed close to the black hole and have been tidally disrupted by it (see chapter 4, Problem 7). We therefore expect these black holes to be accreting matter. Although the sources of accreted matter are different from those in interacting binary systems, which we studied in section 4.6, the physical arguments concerning the accretion process are the same. We can therefore expect the centers (or "nuclei") of galaxies to also display phenomena similar to those in interacting binaries, but scaled according to the larger accretor masses. For example, recalling the Eddington luminosity (Eq. 4.142),

$$L_E = \frac{4\pi c G M m_p}{\sigma_T} = 1.3 \times 10^{38} \text{ erg s}^{-1} \frac{M}{M_\odot}, \tag{6.54}$$

for $M \sim 10^6 - 10^9 M_\odot$ we can expect nuclear luminosities of up to $\sim 10^{47}$ erg s^{-1}, outshining an entire large galaxy by many orders of magnitude.

Indeed, some 1–10% of large galaxies can be classified as having **active galactic nuclei** (AGN), which can be defined as nuclei with a significant amount of luminosity that is of nonstellar origin. The range in the above fraction is, to some degree, semantic—it depends on what is defined as a large galaxy, what kind of galaxies one examines, and what is considered a significant amount of nonstellar activity.

There is much evidence for nuclear activity that is not produced by stars. Some or all of the following phenomena are seen in different AGN: large luminosities emerging from compact regions that are unresolved by telescopes, implying sizes less then a few pc; this luminous energy has a spectral distribution that is distinct from that produced by stellar populations, namely, large luminosities at radio, X-ray, and sometimes gamma-ray frequencies; jets of material emerging from the nucleus, often at relativistic speeds, and emitting synchrotron radiation (these jets are thought to be one possible source of the highest energy cosmic rays detected on Earth); emission lines in the IR to X-ray range, with a spectrum indicating excitation of surrounding gas by a powerful source with a nonstellar spectrum; the emission lines are Doppler-broadened to velocities of up to tens of thousands of km s^{-1}, suggesting motion in a deep gravitational well; and large-amplitude variability in luminosity at all wavelengths, and on all timescales, from minutes to decades. Figures 6.17–6.20 show examples of some of these phenomena.

Each of the above phenomena is difficult to explain by processes other than accretion onto a central supermassive black hole. For example, we could invoke a large population of X-ray binaries to explain the nonstellar spectra, but we would then be hard pressed to understand how all these individual systems could vary in phase to produce the large amplitudes of variability. To be more quantitative, from considerations of causality, a significant change in luminosity (say, a doubling) on a timescale of $\Delta t \sim 1$ hr implies a size less than 1 light hour:

$$R < c\Delta t = 3 \times 10^{10} \text{ cm s}^{-1} \times 3600 \text{ s} = 1.1 \times 10^{14} \text{ cm} = 7 \text{ AU}. \tag{6.55}$$

Thus, a huge luminosity, comparable to, or outshining, that of an entire galaxy, is produced in a region smaller than the Solar System.

As another example, some AGN display long and spatially continuous radio jets of material moving at relativistic speeds ($\sim 0.1c$–c), with lengths sometimes reaching ~ 1 Mpc $= 3 \times 10^6$ ly. This requires that the nucleus has been continuously active for

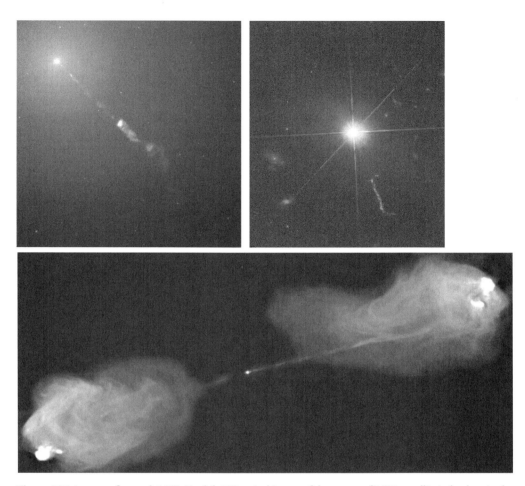

Figure 6.17 Images of several AGN. *Top left:* UV-optical image of the center of M87, an elliptical galaxy in the nearby Virgo cluster of galaxies. M87 contains a relatively low-luminosity active nucleus, evidenced here by the bright unresolved central source, and the jet of material emerging from the nucleus at relativistic speeds. The numerous point sources are globular clusters. *Top right:* Optical image of the quasar 3C273. The light from the nucleus overwhelms the light from the host galaxy, which is barely visible, surrounding the nucleus. A jet can be seen in the lower right quadrant. (The spikes energing from the quasar are artifacts due to diffraction.) *Bottom:* Radio image at 6 cm wavelength of the galaxy Cygnus-A, which has a double-sided jet extending 100 kpc in each direction. Photo credits: NASA and the Hubble Heritage Team; NASA and J. Bahcall; NRAO/AUI, R. Perley, C. Carilli, and J. Dreher.

$\tau \sim 3 \times 10^7$ yr, and therefore has produced a total energy $L\tau$. This energy must equal the radiative efficiency times the rest energy of the mass M involved in the process:

$$E_{\text{tot}} \sim L\tau = \eta M c^2. \tag{6.56}$$

Therefore,

$$M \sim \frac{L\tau}{\eta c^2} = \frac{10^{47} \text{ erg s}^{-1} \times 3 \times 10^7 \times 3.15 \times 10^7 \text{ s}}{\eta \, (3 \times 10^{10} \text{ cm s}^{-1})^2} = \frac{5 \times 10^7 M_\odot}{\eta}. \tag{6.57}$$

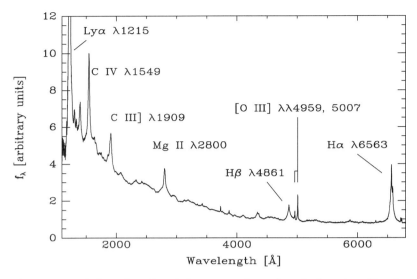

Figure 6.18 Typical UV-through-optical spectrum of an active galactic nucleus. Note the blue continuum, topped by broad emission lines, with the main ones indicated. The emission lines are most probably Doppler broadened by the bulk velocities, 10^3–10^4 km s^{-1}, of the emitting gas clouds in the potential of the supermassive central black hole. Data credit: D. Vanden Berk et al. 2001, *Astron. J.*, 122, 549.

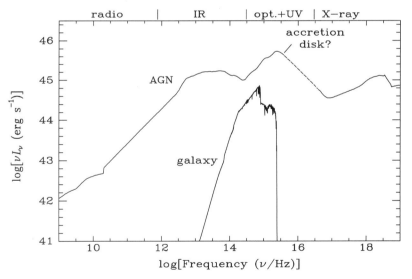

Figure 6.19 Typical spectral energy distribution of an active galactic nucleus, from radio to X-ray frequencies. The vertical axis shows $\nu L_\nu(\nu)$, which has units of luminosity, and which is convenient for visualizing the relative luminosities in different frequency bands. The much-narrower spectral energy distribution of stars in a typical galaxy is shown for comparison. The indicated "bump" in the optical-to-UV range in the AGN spectral distribution may be due to thermal radiation from an accretion disk around the central black hole. The dashed section of the curve is an interpolation in the extreme-UV band, where observations are difficult, because of strong absorption of light from extragalactic sources by neutral hydrogen in the interstellar medium of the Milky Way. Data credit: M. Elvis et al. 1994, *Astrophys. J. Suppl.*, 95, 1.

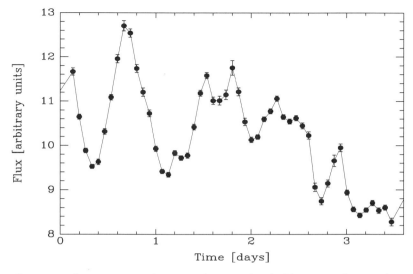

Figure 6.20 Flux variations in the 2- to 10-keV X-ray band of the active galactic nucleus NGC 3516, over several days. Note the large variation amplitudes on timescales of hours, and the nonperiodic, random nature of the variability. Data credit: A. Markowitz and R. Edelson 2005, *Astrophys. J.*, 617, 939.

If AGN were somehow powered through stellar processes by a continuous supply of material undergoing nuclear reactions, which have $\eta = 0.007$ or less, the mass required would be at least $M \sim 10^{10} M_\odot$, comparable to the stellar mass of an entire galaxy. On the other hand, if the energy source is accretion onto a black hole, with $\eta = 0.057$, the required mass is $M \sim 10^9 M_\odot$. This is just the mass of the compact central objects (presumably black holes) found in the nuclei of some massive nearby galaxies.

AGN have been classified into a large zoo of different types, according to the combinations and the details of the properties, listed above, that they display. A few of the common types are Seyfert galaxies, radio galaxies, BL-Lacertae objects, and quasars, and these types are further separated into subcategories (e.g., Seyfert 1 and 2, radio-loud and radio-quiet quasars). The physical distinctions between the different classes and subclasses are still being debated, i.e., what causes the different appearance and properties of different types of AGN. Some of the differences are likely due to orientation effects. The accretion disk, and probably other surrounding AGN components as well, are not spherically symmetric. Their contribution to the total light output and their ability to obscure other components therefore depend on the inclination of the system to the observer's line of sight. It is also yet unclear why the majority of central black holes in galaxies (including the $4 \times 10^6 M_\odot$ black hole at the center of the Milky Way) are not active, and display only feeble signs of nonstellar activity. One possibility is that the gas in the central regions of a galaxy, perhaps by virtue of its high angular momentum or low viscosity, avoids falling into the black hole. Alternatively, perhaps it falls, but in a radiatively inefficient way, and carries most of its kinetic energy with it into the black hole.

The most luminous AGN are **quasars**, roughly defined as AGN with luminosities of 10^{44} erg s^{-1} or more. Among quasars, the most energetic objects reach luminosities of order 10^{47} erg s^{-1}, similar to the Eddington limit estimated above for the largest black-hole masses. Assuming quasars are powered by accretion disks around black holes with inner radii at the last stable orbit, $r_{in} = 3r_s = 6GM/c^2$, we saw (Eq. 4.133) that the luminosity is

$$L \approx \frac{1}{2} \frac{GM\dot{M}}{r_{in}} = \frac{\dot{M}c^2}{12}, \tag{6.58}$$

and, after accounting for general relativistic effects,

$$L = 0.057\dot{M}c^2. \tag{6.59}$$

For a quasar to shine at the Eddington luminosity, its accretion rate $\dot{M}$ must be large enough that

$$L = L_E = 1.3 \times 10^{38} \text{ erg s}^{-1} \frac{M}{M_\odot}, \tag{6.60}$$

and therefore

$$\dot{M} = \frac{1.3 \times 10^{38} \text{ erg s}^{-1} \times 3.15 \times 10^7 \text{ s yr}^{-1}}{0.057c^2} \frac{M}{M_\odot} = 4 \times 10^{-8} M_\odot \text{ yr}^{-1} \frac{M}{M_\odot}. \tag{6.61}$$

Thus, a $10^9 M_\odot$ black hole radiating at the Eddington luminosity must accrete $40 M_\odot$ per year.

The temperature at the inner radius, $r_{in} = 6GM/c^2$, is (Eq. 4.135)

$$T(r_{in}) = \left(\frac{GM\dot{M}}{8\pi\sigma}\right)^{1/4} r_{in}^{-3/4}$$

$$= \frac{c^{3/2}\dot{M}^{1/4}}{6^{3/4}(8\pi\sigma)^{1/4}(GM)^{1/2}} = 3 \times 10^9 \text{ K} \left(\frac{\dot{M}}{M_\odot \text{ yr}^{-1}}\right)^{1/4} \left(\frac{M}{M_\odot}\right)^{-1/2}. \tag{6.62}$$

For the above black-hole mass and accretion rate, the inner temperature will thus be

$$T = 3 \times 10^9 \text{ K} \times 40^{1/4} \times (10^9)^{-1/2} = 2.4 \times 10^5 \text{K}, \tag{6.63}$$

producing thermal radiation that peaks in the extreme UV, between the UV and the X-ray ranges. Quasars, in fact, have such a UV "bump" in their spectral energy distributions (see Fig. 6.19), which is thought to be the signature of the accretion disk.

Since quasars are 10^1–10^4 times brighter than an L_* galaxy with $\sim 10^{10}$ stars, the host galaxies of quasars are often difficult to detect under the glare of the nucleus, especially since the most luminous quasars are generally distant.[6] Owing to the finite speed of light, a large distance means a large lookback time. The large distances of luminous quasars therefore imply that quasars are rare objects at present. Apparently, most central black holes in present-day galaxies are accreting at low or moderate rates, compared to the rates

---

[6] The name "quasars" evolved from the acronym QSRS, for "quasi-stellar radio source"—the first quasars were discovered in the 1960s as radio sources that, at visual wavelengths, have a spatially unresolved, "stellar" appearance. Quasars are often also called QSOs, for "quasi stellar objects." It is more recently, using high-resolution observations in the IR and from space, that the host galaxies of quasars are routinely detected.

that would produce a luminosity of L_E. Quasars were much more common in the past, and their space density reached a peak about 10 Gyr ago. The reasons for this "rise and fall" of the quasar population are unclear. However, there likely is a connection between the growth and development of galaxies and of their central black holes. We will briefly return to these questions in chapter 9.

6.4 Groups and Clusters of Galaxies

Rather than being distributed uniformly in space, galaxies are "spatially correlated"—near a location where there is a galaxy, one is likely to find another galaxy. In other words, galaxies are often found in **groups** consisting of a few to ten members, and sometimes in rich **clusters**, containing up to about 100 luminous (of order L_*) galaxies. We are part of a group of galaxies called the *Local Group*, consisting of the Milky Way, M31 (the Andromeda galaxy), M33, and some smaller "dwarf" galaxies. The nearest rich cluster is the Virgo cluster, at a distance of about 15 Mpc. The typical radius of a galaxy cluster is about $r_{cl} \sim 1$ Mpc. Galaxy clusters have a high proportion of elliptical galaxies, compared to regions with a lower space density of galaxies. In fact, most of the stars in rich clusters are in ellipticals. Figure 6.21 shows optical and X-ray views of a massive galaxy cluster.

Based on the Doppler line-of-sight velocities measured for the individual galaxies in clusters, it is seen that the galaxies in clusters move on orbits with random inclinations and eccentricities, reminiscent of the stars in elliptical galaxies and in spiral bulges. The typical **velocity dispersion**, i.e., the root-mean-square velocity of the cluster galaxies about the mean velocity, is about $\sigma \sim 1000$ km s^{-1}. A typical *cluster crossing timescale*—the time it takes for a galaxy to cross the cluster—is therefore

$$\tau_{cc} \sim \frac{r_{cl}}{\sigma} = \frac{3.1 \times 10^{24} \text{ cm}}{10^8 \text{ cm s}^{-1}} = 3 \times 10^{16} \text{ s} = 10^9 \text{ yr}. \tag{6.64}$$

Several lines of evidence indicate that clusters are old systems, with ages comparable to the age of the Universe, $\sim 10^{10}$ yr. One example of this evidence is the fact that most of the stars are in elliptical galaxies, which consist of old stellar populations.

With a cluster crossing time of a billion years, most galaxies in a cluster have had time to perform a few orbits in the cluster potential, and a cluster is therefore a gravitationally bound system. Note that, with 100 galaxies within a radius of 1 Mpc, the mean galaxy space density in clusters is 10^4 times higher than the average for galaxies in general, and relative velocities between galaxies are a few times higher than between unbound "field" galaxies. The time between collisions (Eq. 6.53) is therefore reduced to a few billion years, and most galaxies in the central regions of clusters have undergone collisions with other galaxies. This may be the reason for the preponderance of ellipticals in clusters.

We can estimate a cluster's **virial mass**, assuming the galaxy velocities trace the cluster potential, as

$$M \sim \frac{\sigma^2 r_{cl}}{G} = \frac{(10^8 \text{ cm s}^{-1})^2 \times 3.1 \times 10^{24} \text{ cm}}{6.7 \times 10^{-8} \text{ cgs}} = 4 \times 10^{47} \text{ g} = 2 \times 10^{14} M_\odot, \tag{6.65}$$

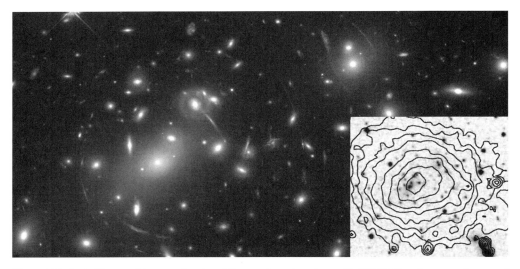

Figure 6.21 Optical image of the central region of the massive galaxy cluster Abell 2218. Note the preponderance of elliptical galaxies. The conspicuous arcs are gravitationally lensed images of distant background galaxies projected near the center of the cluster. The arcs are effectively partial Einstein rings, which allow a measurement on the total mass projected internal to them. Image width is about 0.6 Mpc. *Inset:* Contours showing the X-ray intensity from the cluster, superimposed on an optical image. The area within the inner three contours corresponds approximately to the area shown in the main high-resolution image. The X-ray emission is bremsstrahlung from hot intracluster gas bound by the gravitational potential of the cluster. The gas mass is an order of magnitude greater than the stellar mass of the cluster. The gas temperature provides an independent measure of the total cluster mass. Photo credits: NASA, A. Fruchter, and the ERO Team; and F. Govoni et al. 2004, *Astrophys. J.*, 605, 695, reproduced by permission of the AAS.

where we have used the typical velocity dispersion and radius given above. Galaxy clusters are the most massive bound systems known. An independent estimate of cluster mass can be obtained for many clusters via gravitational lensing. Some rich clusters display large gravitationally lensed arcs (see Fig. 6.21). These are the tangentially stretched images of distant galaxies projected almost directly behind the clusters. When the arcs subtend a substantial angle, they are effectively partial Einstein rings, and their angular radius is approximately the Einstein angle. A typical lensed arc can have a radius of about 0.5 arcmin. Clusters with such arcs are usually at distances of order $D \sim 1$ Gpc, so the Einstein angle corresponds to an **Einstein radius** of

$$R_E = \theta_E D \sim \frac{0.5}{60 \times 180/\pi} 1 \text{ Gpc} = 1.5 \times 10^{-4} \text{ Gpc} = 0.15 \text{ Mpc}. \qquad (6.66)$$

Equation 6.25, which we developed for the Einstein angle of a point mass, also holds for any spherical mass distribution, in which case it gives the total mass projected inside the Einstein ring. Let us take $\theta_E \sim 0.5$ arcmin, and assume again the "halfway" location of the lens (i.e., the cluster) relative to the source (the lensed galaxy), so $D_{ol} = D_{ls} = D_{os}/2$. Then Eq. 6.27 gives for the cluster mass projected within 0.15 Mpc:

$$M \sim \frac{\theta_E^2 c^2 2 D_{ol}}{4G}$$

$$\sim \frac{\left(\frac{0.5}{60 \times 180/\pi}\right)^2 \times (3 \times 10^{10} \text{ cm s}^{-1})^2 \times 10^9 \times 3.1 \times 10^{18} \text{ cm}}{2 \times 6.7 \times 10^{-8} \text{ cgs}}$$

$$= 4 \times 10^{47} \text{ g} = 2 \times 10^{14} M_\odot. \tag{6.67}$$

This is of the same order of magnitude as the virial mass estimate we obtained using the galaxies as kinematic trace particles. (Note, however, that the two mass estimates are measuring somewhat different things; the virial estimate gives the mass within a spherical region of radius 1 Mpc, while the lensing estimate gives the mass enclosed in a long cylindrical volume of radius 0.15 Mpc, with the long axis of the cylinder pointed at the observer.)

Recall now, that a rich cluster consists of about $100 L_*$ galaxies, i.e., a mass of about $10^{13} M_\odot$ in stars. Thus, stars constitute at most a few percent of the mass of a cluster. X-ray observations have revealed that a mass that is about ten times the mass in stars resides in a hot ionized gas of temperature $T \sim (2\text{--}10) \times 10^7$ K (i.e., $kT \sim 2\text{--}10$ keV), which constitutes the **intracluster medium**. The gas is approximately in hydrostatic equilibrium in the cluster potential and radiates thermal bremsstrahlung at X-ray photon energies (see Fig. 6.21, inset). The gas temperature can be derived from the form of the bremsstrahlung spectrum. The intracluster gas provides yet a third estimate of the total mass of a cluster. To see this, assume that the gas has a roughly spherical distribution with radius $r_{gas} \sim 0.5$ Mpc, and recall the equation of hydrostatic equilibrium (Eq. 3.19) from stellar structure, now expressed as an approximate scaling relation:

$$P \sim \frac{GM\rho}{r_{gas}}. \tag{6.68}$$

Note that P is the pressure of the gas, but M is the mass contributing to the potential, i.e., it is the total mass due to all the cluster constituents. Equating to the classical ideal-gas equation of state,

$$P = \frac{\rho}{\bar{m}} kT, \tag{6.69}$$

we obtain

$$M \sim \frac{r_{gas} kT}{G \bar{m}}$$

$$= \frac{0.5 \times 3.1 \times 10^{24} \text{ cm} \times 10^4 \text{ eV} \times 1.6 \times 10^{-12} \text{ erg/eV}}{6.7 \times 10^{-8} \text{ cgs} \times 0.5 \times 1.7 \times 10^{-24} \text{ g}} = 2 \times 10^{14} M_\odot, \tag{6.70}$$

as found via galaxy kinematics and lensing. Here we have assumed the gas has a spatially constant temperature (this is indicated by the X-ray observations) of 10 keV and a fully ionized (appropriate at these temperatures), pure hydrogen, composition.

The total mass in "visible" baryons, including the hot gas and the relatively small amount in stars, still accounts for only about 15% of the total mass of a typical cluster, as determined from galaxy kinematics, lensing, or gas temperature. As in galaxies, the majority of the mass in clusters is therefore also in dark matter of unknown composition.

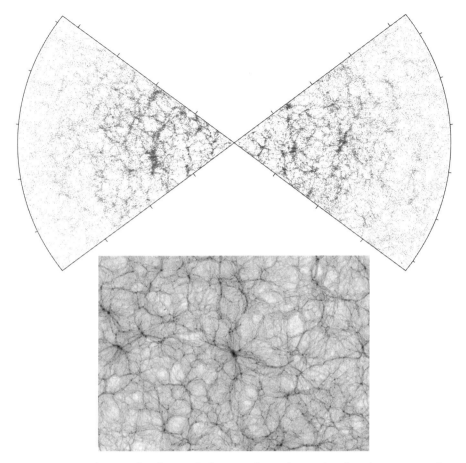

Figure 6.22 *Top:* The space distribution of galaxies, as observed in two thin slices that are centered on our location in the Universe (every dot represents a galaxy). The figure was produced by measuring the two-dimensional positions of the galaxies in two narrow, diametrically opposed, strips on the sky, and the distance to each galaxy. Note the large aggregations of galaxies which surround relatively empty "void" regions. Every tick mark in the radial direction corresponds to ~150 Mpc. The "thinning out" of points beyond ~400 Mpc is an observational effect—at these distances, only the rare, more luminous, galaxies are bright enough to have their distances measured accurately. *Bottom:* A slice through the volume of a numerical N-body simulation in which 10^{10} particles move under the influence of their mutual gravitational attractions. Each particle in the calculation represents a mass of ~$10^9 M_\odot$, and the image width is about 500 Mpc. Note the web-like distribution of particles, which is reminiscent of the observed galaxy distribution in the top figure. Figure credits: The 2dF Galaxy Redshift Survey team, see M. Colless et al. 2001 & 2003, http://arXive.edu/abs/astro-ph/0306581; and V. Springel and the Millenium Simulation team, see *Nature*, 435 (2005), 629.

At scales of 10–100 Mpc, galaxies are still not evenly distributed in space. Attempts to map the distribution of galaxies indicate that galaxies are preferentially distributed along large structures with a variety of morphological descriptions: "walls," "sheets," "filaments," and "bubbles," which surround regions of lower-than-average galaxy density called "voids" (see Fig. 6.22). The emerging picture is of a "cosmic web" or "foam-like" galaxy distribution, with clusters, and aggregates of clusters (called "superclusters") concentrated at the junctions of several bubble walls. Numerical models that attempt to reproduce the formation of these structures via the action of gravity alone have had a fair amount of success (see Fig. 6.22 for an example).

The hierarchy of progressively larger and larger structures ends at ~100 Mpc. Beyond these scales, the Universe appears homogeneous. This means, e.g., that the total mass included within spheres having a 100 Mpc radius is nearly the same at all locations in space. This fact will form one of the foundations for the next chapters.

Problems

1. Even when distances to individual stars are not known, much can be learned simply by counting stars as a function of limiting flux. Suppose that, in our region of the Galaxy, the number density of stars with a particular luminosity L, $n(L)$, is independent of position. Show that the number of such stars observed to have a flux greater than some flux f_0 obeys $N(f > f_0) \propto f_0^{-3/2}$. Explain why the same behavior will occur even if the stars have a distribution of luminosities, as long as that distribution is the same everywhere. If you observed that the numbers do not grow with decreasing f_0 according to this relation, what could be the reason?

   Hint: $N(f > f_0)$ is proportional to the volume of the sphere centered on us, with a radius at which the flux from a star with luminosity L is f_0. The relation between luminosity, flux, and distance then gives the stated result.

2. Derive the expression for gravitational focusing, the increase in the effective cross section for a physical collision between two objects due to their gravitational attraction (Eq. 6.12), as follows. Consider a point mass approaching an object of mass M and radius r_0. When the distance between the two is still large, their relative velocity is v_{ran} and the impact parameter (i.e., the distance of closest passage if they were to continue in relative rectilinear motion) is b. Due to gravitational attraction, the point mass is deflected toward the object and, at closest approach, grazes the object's surface at velocity v_{max}.

   a. Invoke energy conservation to show that $v_{ran}^2 = v_{max}^2 - v_{esc}^2$, where $v_{esc} = (2GM/r_0)^{1/2}$ is the escape velocity from the surface of the star.

   b. Show that angular momentum conservation means that $bv_{ran} = r_0 v_{max}$.

   c. Combine the results of (a) and (b) to prove that gravitational focusing results in an effective cross section for a collision that equals the geometrical cross section of the object times a factor $(1 + v_e^2/v_{ran}^2)$.

3. In the solar neighborhood, the Milky Way has a flat rotation curve, with $v(r) = v_c$, where v_c is a constant, implying a mass density profile $\rho(r) \sim r^{-2}$ (Eq. 6.18).

  a. Assume there is a cutoff radius R, beyond which the mass density is zero. Prove that the velocity of escape from the galaxy from any radius $r < R$ is

$$v_e^2 = 2v_c^2 \left(1 + \ln \frac{R}{r}\right).$$

  b. The largest velocity measured for any star in the solar neighborhood, at $r = 8$ kpc, is 440 km s^{-1}. Assuming that this star is still bound to the galaxy, find a lower limit, in kiloparsecs, to the cutoff radius R, and a lower limit, in units of $M_\odot$, to the mass of the galaxy. The solar rotation velocity is $v_c = 220$ km s^{-1}.
     Answers: $R > e \times 8$ kpc $= 22$ kpc, $M > 2.4 \times 10^{11} M_\odot$.

4. A gravitational lens of mass M is halfway along the line between a source and an observer, at a distance d from each, and produces an Einstein ring. Find the angular radius of the ring, in terms of M and d. For a one-solar-mass lens, at what distance d will the Eintein ring's radius appear larger than a solar radius? (i.e., the ring will not be hidden behind a Sun-like star acting as the lens).
   Answer: 1.6×10^{16} cm $= 0.0052$ pc.

5. We calculated (Eq. 6.41, and below it) the probability that a star in the LMC is gravitationally lensed by MACHOs composing the Galaxy's dark halo. Repeat this calculation, but for star–star lensing in the direction of the Galaxy center, i.e., estimate, to an order of magnitude, the probability that a star in the bulge of our Galaxy will be gravitationally lensed by another star in the galactic disk, close to our line of sight to the bulge star. Assume that we are at a distance $r = 8$ kpc to the bulge, a typical stellar mass is $0.5 M_\odot$, the stellar mass in the disk interior to the solar radius is $5 \times 10^{10} M_\odot$, and the thickness of the disk is $h = 1$ kpc.
   Hint: To simplify the calculation, find the mean number density of stars in the disk, and use this as your "target" density. For the lensing cross section, use the area of a circle formed by the Einstein radius of an $0.5 M_\odot$ star at half the distance to the bulge. The distance covered by the line of sight is our distance to the bulge. The product of the three numbers gives the number of disk stars that are passed to within an Einstein radius by a random line of sight to the bulge. If this number is $\ll 1$, it gives the probability of such an encounter.
   Answer: 2.4×10^{-6}.

6. A point source directly behind a galaxy, at double the observer–galaxy distance d, is gravitationally lensed by the galaxy's potential into an Einstein ring. The galaxy is spherically symmetric, and has a flat rotation curve, $v(r) = v_c$. The bending angle of a light ray passing through a spherical mass distribution with impact parameter r from the distribution's center is

$$\alpha = \frac{4GM(<r)}{c^2 r},$$

   where $M(<r)$ is the mass enclosed within r.

a. Find the angular radius θ_E of the Einstein ring, in terms of v_c, assuming small angles. Calculate θ_E, in arcseconds, for $v_c = 300$ km s^{-1}. At $d = 0.5$ Gpc, to what physical radius, R_E, will this θ_E correspond?

Answers: $\theta_E = 2(v_c/c)^2$; $R_E = 1$ kpc.

b. Distant light sources (e.g., quasars, see also chapter 9), are distributed at random on the sky. It turns out that those of them that are projected behind a galaxy within θ_E of that galaxy will be noticeably lensed. Assume a Euclidean space with a constant number density of galaxies, $n = 10^{-2}$ Mpc^{-3}, R_E from item (a), and a typical distance to the sources of $2d = 1$ Gpc. Write an expression for the fraction of the distant sources that is lensed by intervening galaxies, and evaluate it numerically. Actual surveys that measure the fraction of distant sources that are lensed can probe the properties of the lensing galaxy population, even when those galaxies are not directly observed by means of their light.

Hint: Find the number of "targets" with density n, and with cross-sectional area πR_E^2, that are "hit" by a random line of sight going out to the source distance. Alternatively, consider the fraction of the sky that is covered by the total solid angle within the galaxies' Einstein rings.

Answer: 3.1×10^{-5}.

7. Modified Newtonian dynamics (MoND) proposes that, for small accelerations, Newton's second law, $F = ma$, approaches the form $F = ma^2/a_0$, where a_0 is a constant (see Eq. 6.50).

a. Show how such an acceleration law can lead to flat rotation curves, without the need for dark matter.

b. Alternatively, propose a new law of gravitation to replace $F = GMm/r^2$ at distances greater than some characteristic radius r_0, so as to produce flat rotation curves without dark matter. Make sure your modified law has the right dimensions.

c. Modify further the gravitation law you proposed in (b) with some mathematical formulation (many different formulations are possible), so that the law is Newtonian on scales much smaller than r_0, with a continuous transition to the required behavior at $r \gg r_0$.

7 | Cosmology: Basic Observations

In previous chapters, we have dealt with progressively larger structures. In this chapter, we review the basic observational facts that need to be accounted for by any theory of cosmology—an attempt to describe the nature, history, and future of the Universe as a whole. In chapter 8 such a theory is developed.

7.1 The Olbers Paradox

As we consider more and more distant objects and structures in the Universe, a natural question that emerges is, "Does the Universe have an end?" Due to the finite speed of light, distant objects are seen at large lookback times, leading to another question: "Does the Universe have a finite age, so that light from objects beyond some distance has yet to reach us?" These questions are closely related to the so-called **Olbers paradox**. Olbers's paradox poses the more practical question: "Why is the night sky dark?" The reasoning behind this question is as follows.

Let us assume, naively, that space is Euclidean and that its constituents are static (we will see later that at least the latter assumption is incorrect). Assume furthermore that the Universe is infinite and eternal (i.e., has existed forever). Then every line of sight in the sky must, at some point, reach the photosphere of a star. If f_* is the flux emerging at the star's photosphere, then the flux reaching us from that star will be

$$f_{\text{obs}} = f_* \frac{4\pi r_*^2}{4\pi D_*^2} = f_* \frac{d\Omega}{\pi}, \tag{7.1}$$

where r_* is the stellar radius, D_* is the star's distance, and $d\Omega$ is the solid angle subtended on the sky by the star. Now, since all sightlines eventually hit some stellar surface, the total flux reaching us from all 4π steradians should be

$$f_{\text{tot}} = 4f_*. \tag{7.2}$$

In other words, the whole sky should radiate like a blackbody at a temperature of a few thousand degrees—not only should the night sky be bright, but we would be grilled night and day! Dust along the way does not solve the problem, as every dust particle would also be inside this blackbody radiator and would quickly be heated to the same temperature.

The solution to the Olbers paradox is that at least one of the assumptions must be wrong—the Universe must be finite in extent or in age. To quantify this statement, let us calculate how far we need to look along some random sightline until we encounter a stellar disk. With a mean density of L_* galaxies of about 10^{-2} Mpc^{-3}, and about 10^{10} stars per L_* galaxy, the mean density of stars in the Universe is

$$n_* \sim 10^8 \text{ Mpc}^{-3} = \frac{10^8}{(3.1 \times 10^{24} \text{ cm})^3} \sim 10^{-66} \text{ cm}^{-3}. \tag{7.3}$$

The cross section of a stellar disk is

$$\sigma \sim \pi r_\odot^2 = \pi (7 \times 10^{10} \text{ cm})^2 \sim 10^{22} \text{ cm}^2. \tag{7.4}$$

The mean free path until we "hit" a stellar surface is therefore

$$l = \frac{1}{n_* \sigma} \sim 10^{44} \text{ cm} = 10^{26} \text{ ly}. \tag{7.5}$$

The fact that the sky is dark means that either the size of the Universe

$$l \ll 10^{26} \text{ ly}, \tag{7.6}$$

or the age of the Universe

$$t_0 = \frac{l}{c} \ll 10^{26} \text{ yr}, \tag{7.7}$$

and therefore the light from those stellar surfaces has yet to reach us. The distance beyond which light has yet to reach us is called our **particle horizon**. Actually, the limit is weaker than what we found above, since stars live for a limited time on the main sequence, and then collapse to neutron stars or white dwarfs. However, this does not solve the paradox, as every line of sight would eventually still reach the surface of such a remnant, if not that of a star that is still on the main sequence. Thus, the conclusion that the Universe must be finite in age, extent, or both, still holds.

7.2 Extragalactic Distances

In previous chapters, the distances to, and the sizes of, galactic and extragalactic objects were assumed to be known. In practice, measurements of distance are among the most difficult in astronomy. Broadly speaking, distances are found by many different methods, which comprise a **distance ladder**, going from nearby to distant objects. An overlap in the range of applicability of two different methods allows calibrating the more "distant" method with the more "nearby" method. An obvious shortcoming of such a procedure is that errors in the calibrations of the nearby methods will propagate, and may accumulate to become very large errors in the distant methods. Fortunately, there are a few "direct" methods that skip over several, or all, rungs in the distance ladder, and that serve as a check on the procedure.

Some of the main methods, listed according to the objects and the distance ranges over which they are applicable, are as follows.

a. Nearby stars, within <1 kpc, can have their distances measured by trigonometric parallax, as described in chapter 2. Fairly accurate parallax distances can be obtained currently out to ~100 pc. The luminosities of nearby main-sequence stars with parallax-based distances D_{par} and observed fluxes f can be found from

$$L_{ms} = f \times 4\pi D_{par}^2. \tag{7.8}$$

One can then observe young open star clusters and identify the main sequence in their H-R diagrams. The observed fluxes of the open-cluster main-sequence stars are compared to the known, parallax-based, luminosities of nearby stars of the same spectral type to determine the distances to the clusters:

$$D_{cl} = \sqrt{L_{ms}/(4\pi f)}. \tag{7.9}$$

This technique is called **main-sequence fitting**.

Among the stars in the open clusters there are sometimes pulsating stars called **Cepheids**. Cepheids are a short-lived phase of some intermediate-mass stars during part of the helium-core-burning giant stage. The luminosities L_{cep} of Cepheids with observed fluxes f that are in clusters with main-sequence-fitting distances D_{cl} can again be found from

$$L_{cep} = f \times 4\pi D_{cl}^2. \tag{7.10}$$

The radial pulsations of a Cepheid lead to periodic variations in the luminosity, with a period in the range from about 1 to 100 days, that is a function of the mean luminosity of the star (see Fig. 7.1). One can then calibrate the Cepheid **period-luminosity relation**.[1] Cepheids are our first example of a **standard candle**, a source of known luminosity that can serve as a distance indicator, by comparison of its flux to its intrinsic luminosity.

b. The Large Magellanic Cloud (LMC), Local Group galaxies, and Virgo Cluster galaxies, at <20 Mpc, are near enough that individual luminous Cepheids can be detected, and their periods measured. The deduced Cepheid luminosities can then be compared to their observed fluxes to determine the distances to the galaxies:

$$D_{gal} = \sqrt{L_{cep}/(4\pi f)}. \tag{7.11}$$

In the LMC and in the Local Group one can also find, and measure the parameters of, detached, double-lined, **spectroscopic eclipsing binaries**. Measurement of the orbital velocities of the two components around the center of mass from the *radial velocity curve,*

---

[1] About a dozen Cepheids are near enough to permit direct parallax measurements. These measurements confirm the calibration of the period-luminosity relation.

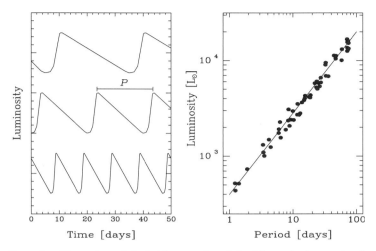

Figure 7.1 Illustration of Cepheid properties. *Left:* Schematic time series for Cepheids of different luminosities. *Right:* Schematic Cepheid period-luminosity relation. Once the relation is calibrated for nearby Cepheids, it can be used to determine the luminosity of Cepheids with measured periods in another galaxy. Comparison of the derived luminosity to the observed Cepheid flux then yields the distance to the galaxy.

and of the durations of the eclipses from the *light curve*, gives the physical radius, r, of each of the two stars (see chapter 2, Problem 5). The observed spectral energy distribution yields the effective temperature, T_E, of each star. The luminosity of each star is then just $L = 4\pi r^2 \sigma T_E^4$, and the combined luminosity of the system can be compared to the observed flux, to derive directly the distance to the system.

Another direct method to obtain the distance to the LMC utilizes a **supernova light echo** that was observed around Supernova 1987A in this galaxy. About 240 days after the explosion, a ring of circumstellar material, ejected by the supernova progenitor star during previous stages in its evolution, became photoionized by the flash of UV radiation from the explosion (see Fig. 7.2). The ring began to shine in emission lines typical of photoionized gas, and is slowly fading as the atoms recombine. The ring is observed to have an angular radius $\theta = 0.85$ arcsecond. The 240-day delay, Δt, was due to the light-travel time from the supernova to the ring, which has a physical radius R:

$$\Delta t = \frac{R}{c} = \frac{\theta D_{\mathrm{lmc}}}{c}. \tag{7.12}$$

(Actually, the ring is inclined to our line of sight, and therefore the front side of the ring was seen to light up earlier than 240 days, and the rear side later, but this geometric effect can be accounted for.) The distance to the LMC is therefore

$$D_{\mathrm{lmc}} = \frac{c\Delta t}{\theta} = \frac{3 \times 10^{10}\ \mathrm{cm\ s^{-1}} \times 240 \times 24 \times 3600\ \mathrm{s}}{0.85/3600 \times \pi/180} = 1.5 \times 10^{23}\ \mathrm{cm} = 50\ \mathrm{kpc}, \tag{7.13}$$

in good agreement with Cepheid-based measurements.

Figure 7.2 The rings around Supernova 1987A. The rings arise in gas that was expelled by the star prior to the supernova explosion. The gas in the bright "equatorial" ring was excited by the flash of UV radiation from the explosion. A delay of 240 days between the explosion and the "lighting up" of the ring was due to the light travel time from the supernova to the ring, indicating a ring radius of 240 light days. Comparison of this physical scale to the angular scale of the ring provides a direct distance measurement to the supernova, and hence to the LMC. Photo credit: C. Burrows, ESA, and NASA.

One Virgo galaxy, NGC 4258, has a direct distance measurement based on the observed **proper motions of water masers** that are moving in circular orbits, with known velocity, around the galaxy's central black hole. See Fig. 7.3 and its caption for details.

c. Galaxies within about 100 Mpc can have their distances estimated by means of several techniques that are calibrated in nearby galaxies (Virgo and closer). The **Tully-Fisher** relation is an empirical correlation between a spiral galaxy's luminosity L and the asymptotic circular velocity, v_c, reached at large radii by its (flat) rotation curve:

$$L \propto v_c^{\alpha}, \tag{7.14}$$

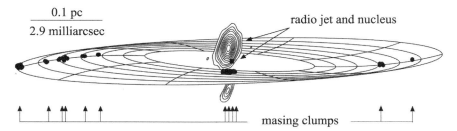

Figure 7.3 Masers in the sub-parsec-scale molecular-gas disk in the center of the galaxy NGC 4258, as observed by radio interferometry using the H_2O 22-GHz emission line. The positions of the masing clumps are shown as they appear on the sky, and superimposed on a model of a nearly edge-on warped disk. The radio (continuum) emission from the inner parts of the jet, emerging from the active nucleus of the galaxy, is also indicated. Based on their measured emission-line Doppler shifts, the velocities of the clumps on the right (approaching) and on the left (receding) follow precisely a Keplerian $r^{-1/2}$ curve, indicating a mass of $3.8 \times 10^7 M_\odot$ for the central black hole. The proper motions of the clumps near the line of sight to the center can be tracked over time as they move from right to left on their circular orbits. At the same time, their centripetal accelerations can be measured through their changing Doppler shifts. Comparison of their angular velocities to their physical velocities gives a direct measurement of the distance to the galaxy. Figure credit: adapted from J. R. Herrnstein, et al. 2005, *Astrophys. J.*, 629, 719, by permission of the AAS.

where the index α depends on the bandpass through which L is measured. Thus, by measuring the rotation velocity of a distant galaxy, one can deduce its luminosity, compare with the observed flux, and derive a distance. Similar relations, called the Faber-Jackson relation, the $D_n\sigma$ relation, and the Fundamental Plane, exist for elliptical galaxies. However, these relations involve the stellar velocity dispersion, rather than v_c—the circular velocity v_c is difficult or impossible to measure directly in most ellipticals, in the absence of a disk of trace particles. **Globular clusters** and **planetary nebulae** in nearby galaxies are observed to have luminosity distributions with a well-defined peak. By measuring the flux distribution of such objects in a distant galaxy, determining the flux corresponding to the peak of the distribution, and comparing to the luminosity of the peak of the distribution in local galaxies, one can deduce the distance.

Another method in this distance regime is the measurement of **surface brightness fluctuations**. Suppose the light from a small solid angle covering part of a galaxy is produced by N stars (see Fig 7.4). The galaxy is distant, and therefore the angular separation between the stars is below the resolution limit of the telescope, and the individual stars cannot be discerned as such. An adjacent small area in the galaxy will include a number of stars that is similar, up to Poisson fluctuations of $\sqrt{N}$. The relative fluctuations in surface brightness (flux per solid angle) measured over a sample of many small regions in a galaxy will therefore be

$$\sigma = \frac{1}{\sqrt{N}}. \tag{7.15}$$

The number of stars included per unit solid angle obviously depends on distance—if a galaxy were moved further away, more stars would be included within the same solid angle, or

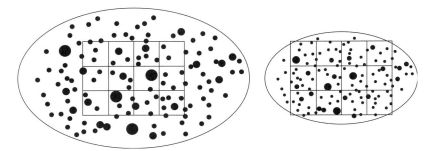

Figure 7.4 Schematic illustration of distance measurement by means of surface brightness fluctuations. On the left, the light output from a small region in a galaxy, e.g., one pixel in the array shown, is contributed by N stars that are not individually resolved as such. The fluctuations in intensity from pixel to pixel will trace the Poisson fluctuations in the number of stars per pixel. On the right, the same galaxy is at a larger distance. As a result, more stars are included in every pixel, and the relative fluctuations, σ, in flux from pixel to pixel will be smaller. Since the amplitude of the relative fluctuations is proportional to distance^{-1} (Eq. 7.17), the fluctuation ratio gives the distance ratio of the galaxies.

$$N \propto D_{\text{gal}}^2.$$ (7.16)

Therefore,

$$\sigma \propto \frac{1}{D_{\text{gal}}}.$$ (7.17)

The proportionality constant can be calibrated in nearby galaxies, and the relation then allows deducing the distance to other galaxies for which surface brightness fluctuations are measured.

d. Galaxies out to $\sim$1 Gpc—Observations of **type Ia supernovae** that have exploded in galaxies that already have distance measurements show that the luminosities of such supernovae at the time of their maximum brightness are approximately the same. A correction to the luminosity based on the rate at which the supernova brightness declines with time makes these supernovae even more reliable standard candles. Measurement of the flux at peak brightness of such a supernova in a distant galaxy then yields the distance to the host galaxy of the event. Owing to the large luminosity of supernovae, this method can be applied to very large distances.

e. Galaxy cluster distances out to $\sim$1 Gpc can be estimated directly via the **Sunyaev-Zeldovich effect**. The hot electrons in the intracluster medium cause *inverse Compton*[2] *scattering*, and therefore a boost in energy, to background photons passing through the cluster

---

[2] Compton scattering, like Thomson scattering, is scattering of photons on electrons, but for the case that the electrons are relativistic, or the photons have energies $> m_e c^2$, or both. With typical intracluster medium temperatures of $kT \sim 10$ keV, the electron velocities are $v \sim (10^{-8} \text{erg}/m_e)^{1/2} \sim 0.1c$, i.e., mildly relativistic. When the electrons are more energetic than the photons, the scattered photons receive an energy boost, and the process is sometimes called inverse Compton scattering.

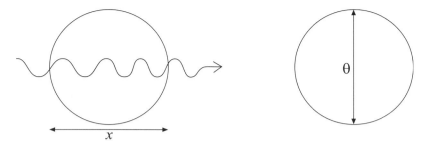

Figure 7.5 Schematic illustration of the principle of galaxy–cluster distance measurement via the Sunyaev-Zeldovich effect. On the left, a photon from the cosmic microwave background passes through the hot gas of the intracluster medium, is slightly boosted in energy through inverse Compton scattering on the electrons in the gas, and continues in its path to the observer. The distortion in the spectral energy distribution of the background photons is proportional, among other things, to the path length x through the cluster. On the right, the emission from the same hot gas, as viewed in X-rays by the observer, subtends an angle θ on the sky. Assuming that the cluster is spherical (and hence its transverse and radial diameters are the same), the distance to the cluster is $D = x/\theta$.

(see Fig. 7.5). A prodigious and well-characterized source of such background photons is the cosmic microwave background, which we will discuss in chapter 9. The energy boost is proportional to the physical path length of the photons through the cluster, and thus the cluster size along the line of sight can be deduced. From the thermal bremsstrahlung X-ray emission of the same hot electron gas, one can also measure an angular size on the sky for the cluster. Comparing the line-of-sight size to the angular size of the cluster, and assuming the cluster is spherical (i.e., not flattened or elongated along our sight line), the distance can be derived (see Problem 3). In reality, clusters are not spherical, but are spherical on average (i.e., they are not all showing us their flat sides or their elongated sides). Therefore, this method can be used to derive the distances to a sample of clusters that are known to be all at about the same distance. The applicability of this will become evident when we learn the concept of cosmological redshift.

7.3 Hubble's Law

The methods outlined above can be used to measure distances to galaxies. The line-of-sight velocity of each galaxy can be measured from the Doppler shift of emission or absorption features in the galaxy spectrum. Since the 1920s, it has been clear that, if one looks at galaxies beyond the nearest ones (e.g., beyond Virgo), all galaxies are receding from us. Furthermore, the recession velocity is linearly proportional to the distance, and follows **Hubble's law**,

$$v = H_0 D. \tag{7.18}$$

The proportionality coefficient is the **Hubble parameter**, $H_0 = 70 \pm 10$ km s^{-1}Mpc^{-1} (often called the *Hubble constant*, though we will see that it is not truly a constant). H_0 has

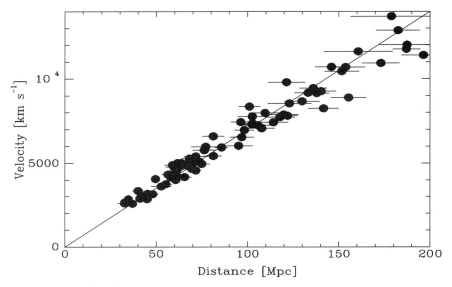

Figure 7.6 A Hubble diagram, showing the recession velocities of galaxies vs. their distances. The distances are deduced from the observed fluxes at maximum light of type Ia supernovae that exploded in the galaxies. The straight line is Hubble's law, $v = H_0 D$, with $H_0 = 70 \, \mathrm{km \, s^{-1} \, Mpc^{-1}}$. Data credit: S. Jha, A. Riess, and R. Kirshner, 2007, *Astrophys. J.*, 659, 122.

dimensions of [velocity]/[distance] = $[\mathrm{time}]^{-1}$. Figure 7.6 shows an example of a "Hubble diagram" displaying this behavior.

Note that the particular type of relative motion embodied by Hubble's law is of a form such that observers living in **any** galaxy see the exact same pattern of recession of the other galaxies surrounding them, with the same proportionality coefficient that we measure. To see this, consider us, at a point O, observing galaxy 1 at vector position $\mathbf{r}_1$ and galaxy 2 at $\mathbf{r}_2$ (see Fig. 7.7). According to Hubble's law, galaxies 1 and 2 have recession velocities

$$\mathbf{v}_1 = H_0 \mathbf{r}_1 \tag{7.19}$$

and

$$\mathbf{v}_2 = H_0 \mathbf{r}_2, \tag{7.20}$$

respectively. Observers on galaxy 2 will, of course, see our galaxy receding from them at velocity $\mathbf{v}_2$, but they will also see galaxy 1 receding from them at velocity

$$\mathbf{v}_{12} = \mathbf{v}_1 - \mathbf{v}_2 = H_0(\mathbf{r}_1 - \mathbf{r}_2) = H_0 \mathbf{r}_{12}, \tag{7.21}$$

which is just Hubble's law from their point of view. (For the vector subtraction of velocities, we have assumed a Euclidean space and nonrelativistic velocities, but the result obtained is general.)

There are several immediate consequences to the observation of Hubble's law. First, all galaxies are moving away from each other, i.e., **the Universe is expanding**. Second, **there is no center** to this expansion since, as we have seen, observers in all galaxies see the same expansion. Third, in the past, the Universe must have been denser. Finally,

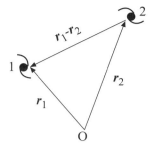

Figure 7.7 An observer at O sees galaxies at vector positions r_1 and r_2 receding according to Hubble's law. Observers in each of those galaxies will see the other two galaxies receding from them according to the same Hubble law.

if the expansion has been going on for long enough, there was a particular time when the distances between all galaxies was zero—this follows from the linear dependence of velocity on distance. Thus, the Universe (at least as we know it) has a finite age. If galaxies do not accelerate or decelerate, then that age, the time since all galaxies were "here," is just

$$t_0 = \frac{1}{H_0} = \frac{1}{70 \text{ km s}^{-1}\text{Mpc}^{-1}} = \frac{3.1 \times 10^{24} \text{ cm}}{70 \times 10^5 \text{ cm s}^{-1}} = 4.4 \times 10^{17} \text{ s} = 14 \text{ Gyr}. \tag{7.22}$$

As we will see in chapter 9, it turns out that this **Hubble time** is close to the current best estimates of the age of the Universe.

7.4 Age of the Universe from Cosmic Clocks

An independent confirmation of the finite age of the Universe comes from several "natural clocks." As we have seen, all elements heavier than oxygen are produced almost solely in stars that undergo supernova explosions. Many of the isotopes produced are radioactive, with known mean lifetimes. Measurement of present-day abundance ratios of various isotopes can reveal when they were produced. For example, from modeling the nuclear reactions that take place during a supernova explosion, it is found that the nuclei of the two isotopes of uranium, ^{235}U and ^{238}U, are produced in similar initial numbers, N_{init}, with a number ratio in the range

$$K \equiv \left[\frac{N_{\text{init}}(^{235}\text{U})}{N_{\text{init}}(^{238}\text{U})} \right]_{\text{SN}} = 1.16\text{–}1.34. \tag{7.23}$$

The 235 isotope has a shorter decay lifetime than the 238 isotope, and therefore the ratio becomes smaller with time. As measured on Earth today, the ratio of the isotopes is

$$K_0 \equiv \left[\frac{N(^{235}\text{U})}{N(^{238}\text{U})} \right]_0 = 0.00723, \tag{7.24}$$

where the zero subscript indicates the present value. The number of nuclei of each isotope decreases from the initial value with time according to

$$N(^{235}\text{U}) = N_{\text{init}}(^{235}\text{U})e^{-t/\tau_{235}}, \tag{7.25}$$

and

$$N(^{238}\text{U}) = N_{\text{init}}(^{238}\text{U})e^{-t/\tau_{238}}, \tag{7.26}$$

where τ_{235} and τ_{238} are the characteristic lifetimes of the two isotopes, which can be measured in the laboratory. Dividing the two equations,

$$K = K_0 \exp\left(\frac{1}{\tau_{238}} - \frac{1}{\tau_{235}}\right) t. \tag{7.27}$$

From the measured values of K_0, τ_{235}, and τ_{238}, the uranium on Earth was produced about 6.2 Gyr ago, if one assumes it was produced all at once from pristine material (that did not contain previously synthesized uranium that had already been decaying). The age resulting from this assumption is, in effect, a lower limit on the age of the Universe; if the Universe had existed for less time, there is no way K_0 could have reached a value as low as observed. A more sophisticated calculation, which takes into account continuous metal enrichment by several generations of star formation and the ensuing supernovae, raises the lower limit to about 10 Gyr.

White dwarf cooling is another cosmic clock. As we saw in section 4.2.3.3, white dwarf temperature is a slow, monotonically decreasing function of time, which detailed models can predict. Thus, the coolest known white dwarfs can provide a lower limit on the age of the Universe. If we assume that stars formed soon after the formation of the Universe, and the first white dwarfs formed shortly thereafter,[3] this is also an estimate (rather than just a lower limit) for the age of the Universe. The coolest white dwarfs give a lower limit of about 10 Gyr for the age of the Universe.

Finally, age dating of globular clusters is possible by identifying the mass corresponding to the main sequence turnoff currently observed in their H-R diagrams (see Fig. 4.1). Stellar structure and evolution models then give the main sequence lifetime of stars of such mass, and this is the age of the cluster. Again, assuming an early formation of globular clusters, this is a lower limit to, or approximately the value of, the age of the Universe. For most globular clusters, ages in the range of 10–15 Gyr are found.

7.5 Isotropy of the Universe

A final, fundamental observation that will guide our development of a cosmological theory is that, at large enough scales, the Universe appears isotropic (meaning "the same in all directions"), i.e., there is no preferred direction in space. This is seen in the distribution on the sky of distant galaxies, as well as of other extragalactic sources, such as gamma-ray

---

[3] Stars of initial mass $\sim 8M_\odot$, at the border between the initial masses leading to either eventual core collapse or to white-dwarf formation, have lifetimes of $\sim$40 Myr.

bursts and quasars. Isotropy is also demonstrated to the extreme by the phenomenon, which we will discuss in chapter 9, called the cosmic microwave background.

Problems

1. Assume that there is a constant ratio between stellar mass and light in the disks of spiral galaxies, and that disks have a constant surface brightness. Use the scaling with radius r, of the stellar mass within such a radius, and the scaling of circular velocity, v_c, with enclosed mass (ignoring dark matter), to "explain" the Tully-Fisher relation, $L \propto v_c^4$. The fourth-power dependence is indeed what is observed at infrared wavelengths, at which light traces stellar mass relatively well. In reality, however, dark matter cannot be ignored, and spiral disks do not have a constant surface brightness.

2. Measurements of the radial recession velocities of five galaxies in a cluster give velocities of 9700, 8600, 8200, 8500, and 10,000 km s^{-1}. What is the distance to the cluster if the Hubble parameter is $H_0 = 70$ km s^{-1}Mpc^{-1}? Estimate, to an order of magnitude, the mass of the cluster if every galaxy is projected roughly half a degree from the cluster center.
   Answers: 130 Mpc; $10^{14} M_{\odot}$.

3. In the Sunyaev-Zeldovich effect, photons from the cosmic microwave background radiation are Compton scattered by hot electrons in a cluster along the line of sight. Assume 0.001 of the photons are scattered, and the mass of the cluster is 2×10^{14} $M_{\odot}$, of which 15% is in the hot gas (fully ionized hydrogen).
   a. Use the Thomson cross section to represent the cross section for Compton scattering, and assume the cluster is spherical and of constant density, to find the diameter of the cluster (assume the photons pass through one diameter).
   b. If the angular diameter of the cluster is 1°, what is its distance?
   c. If the cluster velocity of recession is 8400 km s^{-1}, what is the Hubble parameter, in units of km s^{-1} Mpc^{-1}?
      Answers: diameter 2.1 Mpc; distance 120 Mpc; $H_0 = 70$ km s^{-1}Mpc^{-1}.

8 | Big Bang Cosmology

A successful cosmological theory should reproduce the basic observations outlined in the previous chapter, as well as make testable predictions. In this chapter, we develop such a theory.

8.1 The Friedmann-Robertson-Walker Metric

We will start with the **cosmological principle**, as formulated by Einstein, which postulates that the Universe is **isotropic** and **homogeneous**. The observed isotropy has just been discussed in the previous chapter. Homogeneity is evidenced by the form of Hubble's law, which is such that every observer sees the same expansion. Furthermore, homogeneity takes to the extreme the Copernican principle that we do not hold a special place in the Universe—no one, anywhere in the Universe, holds a special place, and there is full equality of all observers. Note that homogeneity does not necessarily imply isotropy, or vice versa. For example, a spherically symmetric Universe with a radially varying density is not homogenous, but will appear isotropic to an observer at the center. A rotating Universe may be homogeneous, but is not isotropic since it has a preferred direction, along the rotation axis. The requirements that the Universe be isotropic *and* homogeneous is equivalent to the requirement that it appear isotropic from all locations.

The linear form of Hubble's law, $v = H_0 D$, leads (naively, at least) at large enough distances to velocities $v > c$. This suggests that, to describe the dynamics of the Universe, we require a relativistic theory of gravity, namely general relativity. As already discussed in section 4.5, in the context of black holes, general relativity relates the density of mass and energy, which are the sources of gravity, to the curvature of spacetime. The curvature is described by a metric tensor, which specifies the line element of the curved spacetime. Our first task is therefore to find the metric of the Universe that corresponds to the cosmological principle of isotropy and homogeneity.

If space is homogeneous, it must have the same curvature everywhere. There are only three possible geometries that have constant curvature: flat, positively curved, and negatively curved. In two dimensions, these three geometries correspond, respectively, to a plane, the surface of a three-dimensional (3D) sphere, and a surface that at every point has the geometry of a saddle (the latter surface cannot be visualized). All points on the 2D surface of a 3D sphere with radius R (which obviously has constant curvature everywhere, as required) obey

$$x^2 + y^2 + z^2 = R^2, \tag{8.1}$$

and taking the derivative gives

$$xdx + ydy + zdz = 0. \tag{8.2}$$

The line element, giving the distance between two close points, is

$$dl^2 = dx^2 + dy^2 + dz^2 = dx^2 + dy^2 + \frac{(xdx + ydy)^2}{R^2 - x^2 - y^2}. \tag{8.3}$$

Note that the z coordinate is not needed to describe this curved 2D space embedded in a 3D space. In spherical coordinates, the constraint that we must remain on the surface of a sphere simply means that, in the usual 3D line element in spherical coordinates, we set $r = R$ and $dr = 0$, and thus

$$dl^2 = R^2 d\theta^2 + R^2 \sin^2 \theta d\phi^2. \tag{8.4}$$

Note also that no point on this curved 2D surface is preferred, and it has no boundary. It could therefore correspond to a 2D homogeneous and isotropic Universe that is unbounded but finite.

Since we live in a world with three space dimensions, we must extend these concepts to a *hypersphere*, i.e., a positively curved 3D surface, or *3-sphere* of radius R, embedded in a Euclidean 4-space having coordinates x, y, z, and w. The fourth space dimension along the w axis is fictitious and will not be needed to describe the properties of this curved space. In analogy to the 2-sphere,

$$x^2 + y^2 + z^2 + w^2 = R^2, \tag{8.5}$$

and the line element is

$$dl^2 = dx^2 + dy^2 + dz^2 + \frac{(xdx + ydy + zdz)^2}{R^2 - x^2 - y^2 - z^2}. \tag{8.6}$$

Recalling that $x^2 + y^2 + z^2 = r'^2$, where r' is the usual 3D radial coordinate, we can write dl^2 in spherical coordinates as

$$
\begin{aligned}
dl^2 &= dr'^2 + r'^2 d\theta^2 + r'^2 \sin^2 \theta d\phi^2 + \frac{r'^2 dr'^2}{R^2 - r'^2} = \frac{R^2 dr'^2}{R^2 - r'^2} + r'^2 d\theta^2 + r'^2 \sin^2 \theta d\phi^2 \\
&= \frac{dr'^2}{1 - \frac{r'^2}{R^2}} + r'^2 d\theta^2 + r'^2 \sin^2 \theta d\phi^2 \\
&= R^2 \left(\frac{dr'^2 / R^2}{1 - k\frac{r'^2}{R^2}} + \frac{r'^2}{R^2} d\theta^2 + \frac{r'^2}{R^2} \sin^2 \theta d\phi^2 \right).
\end{aligned}
\tag{8.7}
$$

In the last equality we have introduced a *curvature* parameter, k. For the case we have considered, of a hypersphere, $k = +1$. Taking $k = 0$, we recover the usual 3D Euclidean relation, and this corresponds to "flat" 3D space. Taking $k = -1$ gives the line element for a negatively curved 3D space of constant curvature, called a 3-hyperboloid. Finally, if we define a new dimensionless coordinate,

$$r \equiv \frac{r'}{R}, \quad dr \equiv \frac{dr'}{R}, \tag{8.8}$$

and add the time dimension to the line element, we get the spacetime interval between two adjacent events:

$$ds^2 = c^2 dt^2 - dl^2 = c^2 dt^2 - R^2 \left(\frac{dr^2}{1 - kr^2} + r^2 d\theta^2 + r^2 \sin^2 \theta d\phi^2 \right). \tag{8.9}$$

The coefficients of this interval constitute the **Friedmann-Robertson-Walker (FRW) metric**. The meaning of the time coordinate we have introduced here (i.e., time as measured by whom?) will be elucidated soon.

Note that the factor R that multiplies the dimensionless spatial part of the FRW metric is a scale factor. For example, if $R(t)$ grows with time, every observer sees other points in the Universe receding radially, just as in the observed Hubble expansion of galaxies. Thus, a galaxy at coordinates (r, θ, ϕ) remains at those coordinates, and it is the coordinate system which is "locked" onto the galaxies that expands according to $R(t)$. The coordinates (r, θ, ϕ) are therefore called **comoving coordinates**.

The instantaneous distance from us to a galaxy at coordinate r (as would be measured, e.g., by an imaginary taut running tape measure, with one end held at the galaxy and the other end held by us—this is called the **proper distance**) is

$$l = \int_{r=0}^{r} dl = R(t) \int_0^r \frac{dr}{\sqrt{1 - kr^2}} = \begin{cases} R \sin^{-1} r & \text{if } k = +1 \\ Rr & \text{if } k = 0 \\ R \sinh^{-1} r & \text{if } k = -1 \end{cases}. \tag{8.10}$$

For $k = +1$, $r = \sin(l/R)$. The coordinate r reaches a maximum of 1 at a proper distance $l = \pi R/2$, and galaxies beyond this point have smaller r, reaching $r = 0$ at $l = \pi R$, which is our **antipode**. If we travel continuously in one direction, we will pass the antipode, and after traversing a distance $2\pi R$, we will come back to the point of origin, facing the same direction. Similarly, the area of a sphere centered on us and passing through a galaxy at coordinate r, which corresponds to a physical radial coordinate $r' = Rr$, is

$$A = 4\pi r'^2 = 4\pi R^2 r^2 = 4\pi R^2 \sin^2 \frac{l}{R}. \tag{8.11}$$

Beyond $l = \pi R/2$, the area of the sphere decreases, and at the antipode at $l = \pi R$, the sphere centered on us and enclosing all the previous spheres has zero area. This geometrical behavior is the 3D analog of traveling in a certain direction on a 2-sphere, or of drawing concentric circles on a 2-sphere—once a circle passes through a point at a distance (as measured on the surface of the sphere) of $l = \pi R$, its circumference is zero, even though it encloses all the previous circles.

Since

$$l = \int_{r=0}^{r} dl = R(t) \int_{0}^{r} \frac{dr}{\sqrt{1 - kr^2}}, \tag{8.12}$$

and r is a comoving coordinate and therefore is independent of time, the velocity of a galaxy at r is

$$v = \dot{l} = \dot{R}(t) \int_{0}^{r} \frac{dr}{\sqrt{1 - kr^2}} = \frac{\dot{R}}{R} l. \tag{8.13}$$

If we identify the ratio $\dot{R}/R \equiv H(t)$ with the Hubble parameter, we recover Hubble's law. Indeed, Hubble's parameter must depend on time, since we saw it is roughly just the reciprocal of the age of the Universe, and the age increases with time. Stated differently, if, e.g., the galaxies used to measure Hubble's law do not accelerate or decelerate, their distances grow linearly with time, and therefore the Hubble parameter $H = v/D$ becomes smaller with time.

8.2 The Friedmann Equations

We have seen that the FRW metric can describe the three possible constant-curvature geometries of an isotropic and homogeneous Universe, and allows for a Hubble-like expansion described by a scale factor $R(t)$. To proceed and obtain the equations of motion that describe the behavior of this scale factor, we need to specify the mass–energy distribution and relate it to the FRW metric through the Einstein equations of general relativity,

$$G_{\mu\nu} = \frac{8\pi G}{c^4} T_{\mu\nu}. \tag{8.14}$$

Let us see, even if only schematically, how this is done.

As we saw in chapter 4 (Eq. 4.109), the spacetime interval ds is determined by the metric tensor $g_{\mu\nu}$ as

$$(ds)^2 = \sum_{\mu,\nu} g_{\mu\nu} dx_\mu dx_\nu. \tag{8.15}$$

The Einstein tensor $G_{\mu\nu}$ is a combination of first and second derivatives of, in this case, the FRW metric tensor, $g_{\mu\nu}$. Note that the matrix representing the FRW metric tensor is diagonal (there are no cross terms in Eq. 8.9), greatly simplifying $G_{\mu\nu}$. The nonzero terms of $g_{\mu\nu}$ are

$$g_{00} = 1, \quad g_{11} = -\frac{R^2}{1 - kr^2}, \quad g_{22} = -R^2 r^2, \quad g_{33} = -R^2 r^2 \sin^2 \theta. \tag{8.16}$$

To find the Eintein tensor $G_{\mu\nu}$ starting from $g_{\mu\nu}$ (readers unfamiliar with tensor calculus can skip down to Eq. 8.22), one needs to calculate the *affine connections*,

$$\Gamma^{\mu}_{\sigma v} = \frac{1}{2}g^{\mu\rho}\left(\frac{\partial g_{\sigma\rho}}{\partial x^{v}} + \frac{\partial g_{v\rho}}{\partial x^{\sigma}} - \frac{\partial g_{\sigma v}}{\partial x^{\rho}}\right);$$

(8.17)

the Riemann tensor,

$$\mathcal{R}^{\alpha}_{\beta\gamma\delta} = \frac{\partial \Gamma^{\alpha}_{\beta\delta}}{\partial x^{\gamma}} - \frac{\partial \Gamma^{\alpha}_{\beta\gamma}}{\partial x^{\delta}} + \Gamma^{\alpha}_{\rho\gamma}\Gamma^{\rho}_{\beta\delta} - \Gamma^{\alpha}_{\rho\delta}\Gamma^{\rho}_{\beta\gamma};$$

(8.18)

the Ricci tensor,

$$\mathcal{R}_{\beta\gamma} = g^{\alpha\delta}\mathcal{R}^{\alpha}_{\beta\gamma\delta};$$

(8.19)

the Ricci scalar,

$$\mathcal{R} = g^{\beta\gamma}\mathcal{R}_{\beta\gamma},$$

(8.20)

and finally the Einstein tensor,

$$G_{\mu v} = \mathcal{R}_{\mu v} - \tfrac{1}{2}g_{\mu v}\mathcal{R}.$$

(8.21)

In all these relations, the index summation convention is implied, in which summation over all four coordinates is carried out whenever the same index appears jointly as a subscript and a superscript. The first two diagonal components of the Einstein tensor are

$$G_{00} = \frac{3}{c^2 R^2}(\dot{R}^2 + kc^2), \quad G_{11} = -\frac{2R\ddot{R} + \dot{R}^2 + k}{c^2(1 - kr^2)}.$$

(8.22)

In general, the components of the energy-momentum tensor $T_{\mu v}$, which is always symmetric (or can be symmetrized), are

T_{00} = energy density
T_{0i} = momentum flux
T_{ii} = isotropic pressure
T_{ij} = anisotropic pressure (stress and strain)

where the 0 index refers to the time coordinate and the indices i, j to the three spatial coordinates. For an isotropic and homogeneous Universe, $T_{\mu v}$ is diagonal and

$$T_{\mu\mu} = (P + \rho c^2)\frac{v_{\mu}v_{\mu}}{c^2} - Pg_{\mu\mu},$$

(8.23)

where v_{μ} is the 4-velocity, P is the pressure, and ρc^2 is the mass–energy density. Furthermore, for a comoving observer, $v = (c, 0, 0, 0)$, and therefore

$$T_{00} = \rho c^2, \quad T_{11} = \frac{PR^2}{1 - kr^2}.$$

(8.24)

Substituting the (0, 0) and (1, 1) components of $G_{\mu v}$ and $T_{\mu v}$ into the Einstein equations gives the two equations:

$$\frac{\dot{R}^2 + kc^2}{R^2} = \frac{8\pi}{3}G\rho,$$

(8.25)

and

$$\frac{2\ddot{R}}{R} + \frac{\dot{R}^2 + kc^2}{R^2} = -\frac{8\pi}{c^2} GP. \tag{8.26}$$

(The equations resulting from the two other nonzero components of $G_{\mu\nu}$ and $T_{\mu\nu}$ are redundant with these equations. This is a consequence of the isotropy inherent to the FRW metric.) Subtracting the two equations from each other, and slightly rearranging the first equation, gives the first and second **Friedmann equations**,[1] which relate the first and second time derivatives of the scale factor R to the energy density, pressure, and curvature of the Universe:

$$\left(\frac{\dot{R}}{R}\right)^2 = \frac{8\pi}{3} G\rho - \frac{kc^2}{R^2}, \tag{8.27}$$

and

$$\frac{\ddot{R}}{R} = -\frac{4\pi G}{3c^2}(\rho c^2 + 3P). \tag{8.28}$$

The Friedmann equations are two coupled differential equations for the three unknown functions $R(t)$, $\rho(t)$, and $P(t)$. Given an equation of state, $P(\rho)$, and suitable boundary conditions, they can be solved. However, even before we solve them, some immediate consequences are apparent. The first consequence of the Friedmann equations is that the Universe must be expanding or contracting. We know that the Universe has some nonzero mass density ρ, and therefore Eq. 8.27 tells us that $\dot{R}$ is nonzero. (Even if the right-hand side of Eq. 8.27 is momentarily zero, Eq. 8.28 guarantees that this is only momentary.) Equation 8.28 says the acceleration is always negative, i.e., the Universe is decelerating, and always has been. Since the Hubble law shows us that the Universe is currently expanding, the Universe was expanding in the past too, and even faster than now.[2] Thus, if the assumptions of homogeneity and isotropy are valid, and the formulation of general relativity that we have presented is correct, it is unavoidable that the Universe began in an infinitely dense state not more than 14 Gyr ago, the **Big Bang**, and has been expanding since.

The two Friedmann equations imply a third useful relation, obtained as follows. Equation 8.27, slightly rearranged, is,

$$\dot{R}^2 = \frac{8\pi}{3} G\rho R^2 - kc^2. \tag{8.29}$$

Taking the time derivative of both sides gives

$$2\dot{R}\ddot{R} = \frac{8\pi}{3} G\rho 2R\dot{R} + \frac{8\pi}{3} G\dot{\rho}R^2. \tag{8.30}$$

---

[1] The second equation is often called the *acceleration equation.*

[2] In section 8.5, we will find a more general formulation of the Friedmann equations that actually does allow for positive acceleration of R. In chapter 9, we will see that such an acceleration, under the influence of a yet-unexplained form of "dark energy," is likely taking place now, and probably also occurred in the very early Universe.

Substituting Eq. 8.28 for $\ddot{R}$ we get

$$2\dot{R}\left[-\frac{4\pi GR}{3c^2}(\rho c^2 + 3P)\right] = \frac{8\pi}{3}G\rho 2R\dot{R} + \frac{8\pi}{3}G\dot{\rho}R^2. \tag{8.31}$$

This simplifies to

$$-\dot{R}(\rho c^2 + 3P) = 2\rho c^2\dot{R} + \dot{\rho}c^2R. \tag{8.32}$$

Collecting like terms gives

$$-3\dot{R}(\rho c^2 + P) = \dot{\rho}c^2R, \tag{8.33}$$

or

$$\dot{\rho}c^2 = -3\frac{\dot{R}}{R}(\rho c^2 + P), \tag{8.34}$$

which, as we will see, expresses the conservation of energy, and which is often called the *third Friedmann equation*, the *fluid equation*, or the *energy conservation equation*. Note that, in the three equations, ρ is a mass density and ρc^2 is an energy density. Thus, if the dominant source of energy density is not the rest mass density (e.g., if it is mainly a radiation density, $\rho_{\rm rad}$), then we will replace ρ with $\rho_{\rm rad}/c^2$ in each equation.

8.3 History and Future of the Universe

Solving the Friedmann equations for $R(t)$ can give a description of the history and future of the Universe. First, however, an equation of state, $P(\rho)$, needs to be specified. There are two important cases. In the **matter-dominated** case, the pressure from all sources is much less than the matter density,

$$P \ll \rho c^2. \tag{8.35}$$

Setting $P = 0$ in Eq. 8.34, we find

$$\frac{\dot{\rho}}{\rho} = -3\frac{\dot{R}}{R}, \tag{8.36}$$

which has the solution

$$\rho \propto R^{-3}. \tag{8.37}$$

A second important case is when the dominant energy density comes from ultrarelativistic particles (e.g., photons), which have a pressure that is 1/3 of their energy density (see Eq. 3.74),

$$P = \tfrac{1}{3}u = \tfrac{1}{3}\rho c^2. \tag{8.38}$$

In this **radiation-dominated** case, Eq. 8.34 becomes

$$\frac{\dot{\rho}}{\rho} = -4\frac{\dot{R}}{R}, \tag{8.39}$$

with the solution

$$\rho \propto R^{-4}. \tag{8.40}$$

Let us consider now the history of the scale factor $R(t)$. Since ρ behaves as R^{-3} to R^{-4}, in Eq. 8.27 one can always find an early enough time, when R was small enough, such that the second term, which goes as R^{-2} can be neglected,

$$\frac{8\pi}{3}G\rho \gg \left|\frac{kc^2}{R^2}\right|. \tag{8.41}$$

In the matter-dominated era, when $\rho \sim R^{-3}$, Eq. 8.27 then becomes approximately

$$\left(\frac{\dot{R}}{R}\right)^2 \sim \frac{1}{R^3}, \tag{8.42}$$

or

$$R^{1/2}dR \sim dt. \tag{8.43}$$

Integration gives

$$R^{3/2} \propto t, \tag{8.44}$$

and

$$R(t) \propto t^{2/3}. \tag{8.45}$$

Since the energy density of radiation falls faster with R than that of matter, at an early enough time there must have been a radiation dominated era, during which $\rho \sim R^{-4}$. At that time

$$\left(\frac{\dot{R}}{R}\right)^2 \sim \frac{1}{R^4}, \tag{8.46}$$

or

$$R\,dR \propto dt, \tag{8.47}$$
$$R^2 \propto t, \tag{8.48}$$

and

$$R(t) \propto t^{1/2}. \tag{8.49}$$

Note that the reason why the expansion is slower during the radiation-dominated phase, compared to the matter dominated phase ($R \propto t^{1/2}$ vs. $R \propto t^{2/3}$, respectively) is because the gravitating effect of the radiation pressure, in the former case, contributes to slowing down the expansion.

As we look back to the earliest times, as expected, we find that

$$\lim_{t\to 0} R(t) = 0, \tag{8.50}$$

and

$$\lim_{t \to 0} \rho(t) = \lim_{t \to 0} R^{-4} = \infty. \tag{8.51}$$

This singularity in density as $t \to 0$ is the Big Bang.[3] By now we can also see what is the meaning of the time coordinate t. It is simply a universal, or "cosmic," time that can be measured by all comoving observers since the Big Bang. Since the Universe is homogeneous and isotropic, all comoving clocks advance at the same rate. All observers could, in principle, synchronize their clocks by, e.g., agreeing that $t = t_0$ will occur when the local mean density measured by an observer reaches a particular value, ρ_0.

Looking to the future, at some point we can no longer ignore the curvature term in Eq. 8.27. We then need to consider separately the three possibilities, $k = 0, \pm 1$. If space is flat ($k = 0$), Eq. 8.27 becomes

$$\left(\frac{\dot{R}}{R}\right)^2 = \frac{8\pi}{3} G\rho. \tag{8.52}$$

Recalling the definition of the time-dependent Hubble parameter as

$$H \equiv \frac{\dot{R}}{R}, \tag{8.53}$$

this can be rewritten as

$$\rho = \frac{3H^2}{8\pi G}. \tag{8.54}$$

In other words, a flat Universe implies a particular **critical density** for every moment, including now. We previously saw (Eq. 7.22) that, for a current value of the Hubble parameter, $H_0 = 70$ km s^{-1} Mpc^{-1}, $1/H_0 = 4.4 \times 10^{17}$ s, and therefore in cgs units $H_0 = 2.3 \times 10^{-18}$ s^{-1}. At present, the critical density is

$$\boxed{\rho_{c,0} = \frac{3H_0^2}{8\pi G}} \tag{8.55}$$

$$= \frac{3(2.3 \times 10^{-18} \text{ s}^{-1})^2}{8\pi \times 6.7 \times 10^{-8} \text{ cgs}} = 9.2 \times 10^{-30} \text{ g cm}^{-3} = 1.4 \times 10^{11} M_\odot \text{ Mpc}^{-3}.$$

Recall that the typical density of L_* galaxies (see section 6.2) is 10^{-2} Mpc^{-3}, and that such galaxies, including their dark halos, have masses of $\sim 10^{12} M_\odot$. The matter density due to galaxies is therefore of order 10 times less than the critical density. It is convenient to express the actual matter density of the Universe in units of the critical density, by means of the parameter

$$\Omega_m \equiv \frac{\rho}{\rho_c}. \tag{8.56}$$

---

[3] It is a common misconception that, as $t \to 0$, all the matter in the Universe "was concentrated in a single point." This would imply that the Big Bang occurred at a particular location in space, contrary to the cosmological principle. In fact, the singularity is in the density, and it occurs everywhere in the Universe at once. Even if the Universe is infinite ($k = 0$ and $k = -1$ cases), and hence has infinite volume and mass, then no matter how highly compressed is the matter within some volume, there is infinitely more matter outside the volume, and an equally high density is achievable everywhere else.

If $\Omega_m = 1$, then $k = 0$. If, furthermore, we are now in a matter-dominated era (as we will see below that we are), then the approximate solution of Eq. 8.27 that we found before is exact, $R \propto t^{2/3}$, and therefore $\dot{R} \propto t^{-1/3}$. The scale factor (and hence the distance between any two galaxies) continues to grow forever while gradually slowing down, stopping only at $t = \infty$.

Proceeding to the fate of the Universe in the second case, of a positively curved space with $k = +1$, as R continues to grow and the density ρ goes down, there will come a time when the two terms on the right-hand side of Eq. 8.27 are equal,

$$\frac{8\pi}{3} G\rho = \frac{kc^2}{R^2}. \tag{8.57}$$

This will happen when the scale radius

$$R = \left(\frac{3c^2}{8\pi G\rho} \right)^{1/2}, \tag{8.58}$$

at which time

$$\left(\frac{\dot{R}}{R} \right)^2 = 0, \tag{8.59}$$

i.e., the expansion will halt. However, the deceleration in Eq. 8.28 does not change its negative sign, and therefore this is the beginning of a collapse, in which the Universe traces in reverse its past expansion, up to a "Big Crunch." Recall that, since such a recollapsing Universe has positive curvature, its volume is finite but unbounded. It is called a **closed** Universe.

Finally, if $k = -1$, after sufficient time the curvature term will dominate over the density term in Eq. 8.27, so that

$$\left(\frac{\dot{R}}{R} \right)^2 \sim \frac{c^2}{R^2}, \tag{8.60}$$

or

$$\dot{R} = c. \tag{8.61}$$

In other words, the expansion continues forever at a constant, "coasting," rate. The $k = -1$ universe is an **open** universe that is infinite and forever expanding (as is the $k = 0$ case). Figure 8.1 shows examples of the time dependence of $R(t)$ for each of the three curvature possibilities.

When is the transition from a radiation dominated Universe, with $\rho \sim R^{-4}$, to a matter dominated Universe, with $\rho \sim R^{-3}$? The energy density in radiation at any time, ρ_{rad}, is related to its value today, $\rho_{\mathrm{rad},0}$, by

$$\rho_{\mathrm{rad}} = \rho_{\mathrm{rad},0} \frac{R_0^4}{R^4}. \tag{8.62}$$

Similarly for the mass energy density,

$$\rho_{\mathrm{m}} c^2 = \rho_{\mathrm{m},0} c^2 \frac{R_0^3}{R^3}. \tag{8.63}$$

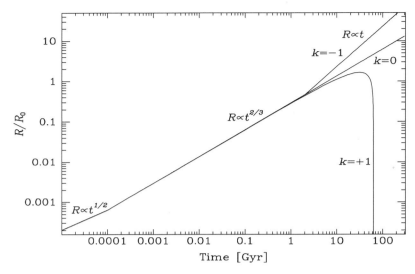

Figure 8.1 Examples of the time dependence of the relative scale factor, R/R_0, for various cosmologies. At early, radiation-dominated times, R grows as $t^{1/2}$, and during early matter domination, $R \sim t^{2/3}$ for all models. The $t^{2/3}$ behavior continues forever in the flat ($k = 0$), critical-density model. In the open, curvature-dominated ($k = -1$) model, the Universe reaches a final "coasting" phase, with $R \sim t$. In the supercritical, $k = +1$, case, R attains a maximum and the Universe then recollapses to a singularity, retracing its past evolution symmetrically in reverse. The asymmetric appearance of the $k = +1$ curve is due to the logarithmic scale of the plot (the logarithmic scale is useful for visualizing the early-time behavior and the various power-law dependences).

The two energy densities were equal when

$$\frac{R_0}{R} = \frac{\rho_{m,0}c^2}{\rho_{rad,0}}. \tag{8.64}$$

As we will see, the Universe is filled today with a radiation field, the cosmic microwave background, that has the spectrum of a blackbody at a temperature $T_0 = 2.73$ K. The radiation energy density today is therefore

$$\rho_{rad,0} = aT_0^4 = 7.6 \times 10^{-15} \text{ erg cm}^{-3} \text{ K}^{-4} \times (2.73 \text{ K})^4$$
$$= 4.2 \times 10^{-13} \text{ erg cm}^{-3}. \tag{8.65}$$

We saw that the present-day matter density is not far (within an order of magnitude) from the critical closure density ρ_c. We will see later on that it is actually about 0.3 of the critical density. Thus,

$$\rho_{m,0}c^2 \approx 0.3\rho_{c,0}c^2 = 0.3 \times 9.2 \times 10^{-30} \text{ g cm}^{-3} \times (3 \times 10^{10} \text{ cm s}^{-1})^2$$
$$= 2.5 \times 10^{-9} \text{ erg cm}^{-3}. \tag{8.66}$$

Therefore, today we are clearly in a matter-dominated era. The transition from domination by relativistic particles to the present, matter-dominated, era occurred when the scale factor R was smaller than its present value R_0 by

$$\frac{R_0}{R} = \frac{\rho_{m,0}c^2}{1.7\rho_{rad,0}} = \frac{2.5 \times 10^{-9} \text{ erg cm}^{-3}}{1.7 \times 4.2 \times 10^{-13} \text{ erg cm}^{-3}} = 3500. \tag{8.67}$$

(The factor 1.7 accounts for the energy density due to the **cosmic neutrino background**, another component of the Universe that must exist, although it has not yet been detected—see chapter 9, Problem 9. At early times these neutrinos, even though they have a nonzero mass, were relativistic, and therefore behaved just like the radiation, with a density proportional to R^{-4}. It can be shown that their energy density then was 0.68 times the photon energy density.) Recall that under matter-dominated conditions R grows as $t^{2/3}$ (Eq. 8.45), and therefore

$$\frac{R_0}{R} = \left(\frac{t_0}{t}\right)^{2/3}. \tag{8.68}$$

Thus, the time of transition to matter domination was

$$t = \frac{t_0}{(3500)^{3/2}} = \frac{t_0}{2 \times 10^5}. \tag{8.69}$$

In other words, the Universal expansion has been matter-dominated for all but a small fraction of the age of the Universe.

To calculate the age of the Universe, we can therefore safely make the approximation that we have had $\rho \sim R^{-3}$ throughout the history of the Universe. Let us examine the various possibilities. First, if $\rho_0 = \rho_{c,0}$, so that $k = 0$, we saw that Eq. 8.27 gives

$$H^2(t) \equiv \left(\frac{\dot{R}}{R}\right)^2 = \frac{8\pi G}{3}\rho \propto R^{-3}. \tag{8.70}$$

Then

$$\left(\frac{\dot{R}}{R}\right)^2 = H_0^2 \frac{R_0^3}{R^3}, \tag{8.71}$$

which after separating the variables and integrating becomes

$$\frac{1}{H_0} \int_0^{R_0} \frac{R^{1/2}dR}{R_0^{3/2}} = \int_0^{t_0} dt, \tag{8.72}$$

or

$$t_0 = \tfrac{2}{3}H_0^{-1}. \tag{8.73}$$

If we consider an empty Universe with $\rho = 0$, which is the extreme case of $k = -1$, Eq. 8.28 becomes

$$\ddot{R} = 0, \tag{8.74}$$

and

$$\dot{R} = \text{const.} \tag{8.75}$$

But

$$H = \frac{\dot{R}}{R},$$
(8.76)

so

$$\dot{R} = HR = H_0 R_0,$$
(8.77)

$$\int_0^{R_0} \frac{dR}{H_0 R_0} = \int_0^{t_0} dt,$$
(8.78)

and

$$t_0 = H_0^{-1}.$$
(8.79)

Thus, for

$$1 > \Omega_{m,0} > 0,$$
(8.80)

the age of the Universe is in the range

$$\tfrac{2}{3} H_0^{-1} < t_0 < H_0^{-1}.$$
(8.81)

We already saw that, for $H_0 = 70 \, \text{km s}^{-1} \text{Mpc}^{-1}$, $H_0^{-1} = 14 \, \text{Gyr}$. The age of the Universe is therefore between 9 and 14 Gyr, for this range of values of the density parameter $\Omega_{m,0}$. The transition from a radiation-dominated to a matter-dominated expansion occurred at a time

$$t = \frac{t_0}{(3500)^{3/2}} = \frac{t_0}{2 \times 10^5} \sim 65,000 \, \text{yr}$$
(8.82)

after the Big Bang, assuming the larger age.

8.4 A Newtonian Derivation of the Friedmann Equations

A more intuitive understanding of the Friedmann equations can be obtained from an approximate derivation based on local Newtonian arguments. Consider a spherical region of radius R, total mass M, and constant density ρ (see Fig. 8.2). A galaxy of mass m is at the edge of the region, at a radius R from an observer at the center. Energy conservation means that

$$\frac{1}{2} m \dot{R}^2 - \frac{GMm}{R} = E,$$
(8.83)

where the total energy, E, is a constant. Replacing M with

$$M = \frac{4\pi}{3} R^3 \rho,$$
(8.84)

we obtain

$$\left(\frac{\dot{R}}{R} \right)^2 = \frac{8\pi G}{3} \rho + \frac{2E}{mR^2}.$$
(8.85)

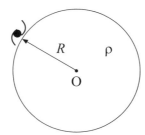

Figure 8.2 A galaxy at the edge of a spherical mass distribution of constant density ρ and radius R, as viewed by an observer at the center. Energy conservation and Newtonian kinematics lead to approximate versions of the Friedmann equations.

By identifying $2E/m$ with $-kc^2$, we recover the first Friedmann equation, Eq. 8.27. We thus see that this equation basically says that the sum of the kinetic and potential energies of the Universe is locally conserved.

The equation of motion for the galaxy, again using Eq. 8.84, is

$$m\ddot{R} = -\frac{GMm}{R^2} = -\frac{4}{3}\pi GR\rho m, \tag{8.86}$$

or

$$\frac{\ddot{R}}{R} = -\frac{4\pi}{3}G\rho. \tag{8.87}$$

This is the second Friedmann equation (Eq. 8.28), except for a missing $3P/c^2$ term—missing since Newtonian gravity does not account for the gravitating effect of pressure. The second Friedmann equation is thus just the equation of motion under the influence of gravity.

The third Friedmann equation (Eq. 8.34) can be obtained from a thermodynamic argument involving only special, rather than general, relativity. Conservation of energy implies that, in a system undergoing adiabatic compression or expansion (i.e., with no net heat flow into or out of the system), the energy U, pressure P, and volume V obey

$$dU = -PdV. \tag{8.88}$$

The adiabatic condition is consistent with the cosmological principle, since in a homogeneous and isotropic Universe there can be no net energy flow from one region to another. Substituting

$$U = \rho c^2 V \tag{8.89}$$

and taking the time derivative on both sides,

$$\frac{d(\rho c^2 V)}{dt} = -P\frac{dV}{dt}, \tag{8.90}$$

we obtain

$$\dot{\rho}c^2 V + \rho c^2\frac{dV}{dt} = -P\frac{dV}{dt}, \tag{8.91}$$

which simplifies to

$$\dot{\rho}c^2 = -\frac{\dot{V}}{V}(\rho c^2 + P). \tag{8.92}$$

Since $V \propto R^3$,

$$\frac{dV}{V} = 3\frac{dR}{R}. \tag{8.93}$$

Substitution in Eq. 8.92 then gives the required result,

$$\dot{\rho}c^2 = -3\frac{\dot{R}}{R}(\rho c^2 + P). \tag{8.94}$$

The third Friedmann equation, which we previously derived by combining the first two equations, is thus basically a restatement of energy conservation. We can also see now that the above derivation holds separately for each of several cospatial systems of particles with no net exchange of energy between the systems (so that the adiabatic condition holds for each system). This can occur if there is no interaction between the systems, or if they are in full thermodynamic equilibrium. For example, in a gas composed of matter and radiation in thermodynamic equilibrium, Eq. 8.94 will hold separately for the matter density and its associated pressure, and for the radiation density and its pressure.

8.5 Dark Energy and the Accelerating Universe

We have derived above the dynamics of a universe that is controlled solely by the gravity due to matter and radiation. However, it is possible, in principle, to add a term, $\Lambda g_{\mu\nu}$, to the Einstein equations, which, we will see, can act as a repulsive force that counteracts the conventional attractive gravity. Such a term, called a **cosmological constant**, was first introduced by Einstein to his equations to allow the existence of a static Universe (which, as we saw, is not possible in the formulation we have developed so far). After the Hubble expansion was discovered, Einstein discarded the cosmological constant, but it has resurfaced several times over the years, by way of attempts to explain a number of different observations. In recent years, evidence is mounting that a Λ-like term may, in fact, be required to describe the dynamics of our Universe. The cosmological constant is one possibility among a class of such terms that can be added to the Einstein equations, which are generally referred to as **dark energy** or **vacuum energy**.

With the addition of Λ, the Friedmann equations that result by writing the Einstein equations for the FRW metric are modified, and become

$$\left(\frac{\dot{R}}{R}\right)^2 = \frac{8\pi}{3}G\rho - \frac{kc^2}{R^2} + \frac{\Lambda}{3} \tag{8.95}$$

and

$$\frac{\ddot{R}}{R} = -\frac{4\pi G}{3c^2}(\rho c^2 + 3P) + \frac{\Lambda}{3}. \tag{8.96}$$

The third Friedmann equation,

$$\dot{\rho}c^2 = -3\frac{\dot{R}}{R}(\rho c^2 + P), \tag{8.97}$$

remains unchanged (see Problem 4).

From Eq. 8.96, it is clear that a large enough positive value of Λ can cause $\ddot{R}$ to become positive, i.e., to make the Universe accelerate, as opposed to the deceleration that always exists without such a term. Note that Λ has dimensions of $[\text{time}]^{-2}$. From Eq. 8.95, we can see that the cosmological constant acts effectively as an additional energy density,

$$\epsilon_\Lambda = \frac{c^2}{8\pi G}\Lambda. \tag{8.98}$$

However, if Λ is constant, ϵ_Λ is an energy density that remains constant, rather than falling, when R grows with time.[4] Thus, after R has grown enough, it is guaranteed that the Λ term will dominate the right-hand side of Eq. 8.95. We can then write Eq. 8.95 as

$$H^2 = \left(\frac{\dot{R}}{R}\right)^2 \approx \frac{\Lambda}{3}, \tag{8.99}$$

or

$$\dot{R} \approx \left(\frac{\Lambda}{3}\right)^{1/2} R, \tag{8.100}$$

which has the solution

$$R(t) \propto \exp\left[\left(\frac{\Lambda}{3}\right)^{1/2} t\right] = \exp(Ht), \tag{8.101}$$

where the Hubble parameter H has actually become a constant. In other words, once the cosmological constant term comes to dominate, the Universe enters an accelerating, exponentially expanding, phase. If Λ remains constant, this phase lasts forever. During the exponential expansion phase, the particle horizon—the most distant point an observer can, in principle, see—tends to a constant comoving coordinate r_h (see Problems 1 and 2). Thus, as opposed to a Universe without a cosmological constant, in which more and more of the volume becomes visible as time progresses, there is a fixed limit beyond which light will never reach us (since, at the time of emission, galaxies beyond that distance are receding from us faster than the speed of light). Galaxies within the particle horizon will get more and more redshifted with time, and therefore an observer in such a universe will see more and more of the light sources around him "blinking out" (actually, getting redshifted to infinity). Finally, observers in an exponentially expanding Universe are surrounded also by an **event horizon**, similar to that around a black hole, that bounds the region of space with which they can communicate or interact causally. The comoving radial coordinate

---

[4] This counterintuitive behavior results from the strange equation of state associated with the cosmological constant, which relates a **negative pressure** to ϵ_Λ: $P = -\epsilon_\Lambda$ (see Problem 5). When a volume element in the Universe grows due to the expansion, the work done by the negative pressure maintains the energy density constant.

of the event horizon, r_{eh}, shrinks exponentially with time, and therefore all observers eventually lose contact with each other (see Problem 3).

It turns out observationally that a model that is particularly relevant to the real Universe is one with a nonzero cosmological constant *and* a flat space. In this case, setting $k = 0$ in Eq. 8.95 and dividing both sides by H_0^2, we obtain

$$\frac{H^2}{H_0^2} = \frac{8\pi}{3H_0^2}G\rho + \frac{\Lambda}{3H_0^2}. \tag{8.102}$$

Recalling our definition of the present critical density for closure,

$$\rho_{c,0} = \frac{3H_0^2}{8\pi G}, \tag{8.103}$$

this can be written as

$$\frac{H^2}{H_0^2} = \frac{\rho}{\rho_{c,0}} + \frac{\Lambda}{3H_0^2}. \tag{8.104}$$

If we recall also the definition of the density parameter,

$$\Omega_m \equiv \frac{\rho}{\rho_c}, \tag{8.105}$$

and define an analogous dimensionless parameter for Λ,

$$\Omega_\Lambda \equiv \frac{\Lambda}{3H_0^2}, \tag{8.106}$$

then Eq. 8.104 *at the present* becomes

$$1 = \Omega_{m,0} + \Omega_{\Lambda,0}. \tag{8.107}$$

However, the same argument can be made at any time, and therefore, if $k = 0$,

$$\boxed{\Omega_m + \Omega_\Lambda = 1} \tag{8.108}$$

always. Thus, as opposed to the flat, zero-Λ, universe, in which the mass density always equals exactly the critical closure density (i.e., $\Omega_m = 1$), in a flat, nonzero-Λ, universe, it is only the *sum* of Ω_m and Ω_Λ that is constant and equal to 1. In a closed, positive-curvature Universe, $\Omega_m + \Omega_\Lambda > 1$, and in an open, negative-curvature Universe, $\Omega_m + \Omega_\Lambda < 1$. In the next chapter, we review recent measurements indicating that $\Omega_m + \Omega_\Lambda$ is very close to 1 (i.e., space is nearly flat/Euclidean). As $k = 0$ marks the border between a closed and an open Universe, it may be difficult to find out whether space has a finite or infinite volume.

Problems

1. Show that the current proper distance to our particle horizon, defined as the most distant place we can see (in principle), for a matter-dominated $k = 0$ universe with no cosmo-

logical constant, is $r_h R_0 = 3ct_0$, where r_h is the comoving radial coordinate of the particle horizon, R_0 is the scale factor today, and t_0 is the present age of the Universe. Thus, more and more distant regions of the Universe "enter the horizon" and become visible as time progresses. Why is the answer different from the naively expected result ct_0?

Hint: Light moves along *null geodesics*, defined as paths along which $ds = 0$, and therefore in the FRW metric, light reaching us from a comoving coordinate r will obey

$$0 = c^2 dt^2 - R(t)^2 \frac{dr^2}{1 - kr^2}.$$

Replace $R(t)$ with $R_0(t/t_0)^{2/3}$ appropriate for this cosmology, separate the variables, and integrate from $r = 0$ to r_h and from $t = 0$ (the Big Bang) to t_0 (today).

2. For a $k = 0$ universe with $\Omega_\Lambda = 1$, that at $t = 0$ already has a scale R_0, find the comoving radial coordinate, r_h, of galaxies that will be on the particle horizon (see Problem 1) at a time t in the future. Show that in this case r_h approaches a constant, $c/(H_0 R_0)$, and therefore galaxies beyond this r_h will never become visible.

   Hint: Proceed as in Problem 1, but now with $R(t) = R_0 \exp(H_0 t)$. (Show why this $R(t)$ is an exact solution of the Friedmann equations for the cosmological parameters above.)

3. a. For the same cosmology as in Problem 2 ($k = 0$, $\Omega_\Lambda = 1$), find the comoving radius r_{eh} of galaxies that will be on our **event horizon** at a time t in the future, i.e., galaxies with which we will be unable to communicate. In other words, light signals sent by us at time t will never reach those galaxies, light signals sent out by those galaxies at time t will never reach us, and therefore we will never see those galaxies as they appeared at time t and thereafter. Show that, in this case, r_{eh} shrinks exponentially, and we thus lose the possibility of communication with more and more of our neighbors.

   b. Assume that $H_0 = 70$ km s^{-1}Mpc^{-1} and find, for such a universe (which approximates the actual world we live in), within how many years will the galaxies in the the nearby Virgo cluster (distance ~ 15 Mpc) reach the event horizon.

   Hint: Proceed as in Problem 2, but integrate from $r = 0$ to r_{eh} and from t (future emission time) to $t = \infty$ (the photons never reach us). Then equate r_{eh} to the comoving radius of Virgo, 15 Mpc/R_0.

   Answer: 79 Gyr.

4. Repeat the derivation of the third Friedmann equation, from the first and second Friedmann equations, but in the presence of a cosmological constant (Eqs. 8.95 and 8.96), and show that this equation is unchanged. Note that, in this derivation, ρ and P still refer to the density and pressure associated with normal matter and radiation, rather than with the cosmological constant term, which cancels out.

5. Show that the equation of state associated with the energy density of the cosmological constant is $P = -\epsilon_\Lambda$, with a negative pressure. Two different ways to do this are as follows.

a. Invoke energy conservation and follow the derivation of Eqs. 8.88–8.94 to argue that Eq. 8.94 holds also for the "dark energy" density component, ϵ_Λ, alone, i.e.,

$$\dot{\epsilon}_\Lambda = -3\frac{\dot{R}}{R}(\epsilon_\Lambda + P_\Lambda).$$

The required result follows from noting that Λ is a constant.

b. Rewrite the Friedmann equations plus cosmological constant (Eqs. 8.95, 8.96), but absorb the $\Lambda/3$ term, i.e., in Eq. 8.95, define an energy density ϵ_Λ (Eq. 8.98) such that ρ is replaced by $\rho + \epsilon_\Lambda/c^2$. In Eq. 8.96, replace $\rho c^2 + 3P$ with $\rho c^2 + \epsilon_\Lambda + 3(P + P_\Lambda)$. Then eliminate Λ from the two defining equations of ϵ_Λ and P_Λ to obtain the required dark energy equation of state.

9 Tests and Probes of Big Bang Cosmology

In this final chapter, we review three experimental predictions of the cosmological model that we developed in chapter 8, and their observational verification. These tests—cosmological redshift, the cosmic microwave background, and nucleosynthesis of the light elements—also provide information on the particular parameters that describe our Universe. We conclude with a brief discussion on the use quasars and other distant objects as cosmological probes.

9.1 Cosmological Redshift and Hubble's Law

Consider light from a galaxy at a comoving radial coordinate r_e. Two wavefronts, emitted at times t_e and $t_e + \Delta t_e$, arrive at Earth at times t_0 and $t_0 + \Delta t_0$, respectively. As already noted in section 4.5 in the context of black holes, the metric of spacetime dictates the trajectories of particles and radiation. Light, in particular, follows a null geodesic with $ds = 0$. Thus, for a photon propagating in the FRW metric (see also chapter 8, Problems 1–3), we can write

$$0 = c^2 dt^2 - R(t)^2 \frac{dr^2}{1 - kr^2}. \tag{9.1}$$

The first wavefront therefore obeys

$$\int_{t_e}^{t_0} \frac{dt}{R(t)} = \frac{1}{c} \int_0^{r_e} \frac{dr}{\sqrt{1 - kr^2}}, \tag{9.2}$$

and the second wavefront

$$\int_{t_e + \Delta t_e}^{t_0 + \Delta t_0} \frac{dt}{R(t)} = \frac{1}{c} \int_0^{r_e} \frac{dr}{\sqrt{1 - kr^2}}. \tag{9.3}$$

Since r_e is comoving, the right-hand sides of both equalities are independent of time, and therefore equal. Equating the two left-hand sides, we find

$$\int_{t_e+\Delta t_e}^{t_0+\Delta t_0} \frac{dt}{R(t)} - \int_{t_e}^{t_0} \frac{dt}{R(t)} = 0. \tag{9.4}$$

Expressing the first integral as the sum and difference of three integrals, we can write

$$\int_{t_e}^{t_0} - \int_{t_e}^{t_e+\Delta t_e} + \int_{t_0}^{t_0+\Delta t_0} - \int_{t_e}^{t_0} = 0, \tag{9.5}$$

and the first and fourth terms cancel out. Since the time interval between emission of consecutive wavefronts, as well as the interval between their reception, is very short compared to the dynamical timescale of the Universe ($\sim 10^{-15}$ s for visual light, vs. $\sim 10^{17}$ s for a Hubble time), we can assume that $R(t)$ is constant between the two emission events and between the two reception events. We can then safely approximate the integrals with products,

$$\frac{\Delta t_e}{R(t_e)} = \frac{\Delta t_0}{R(t_0)}. \tag{9.6}$$

Recalling that

$$\Delta t_e = \frac{1}{\nu_e} = \frac{\lambda_e}{c} \tag{9.7}$$

and

$$\Delta t_0 = \frac{1}{\nu_0} = \frac{\lambda_0}{c}, \tag{9.8}$$

we find that

$$\frac{\Delta t_0}{\Delta t_e} = \frac{\lambda_0}{\lambda_e} = \frac{\nu_e}{\nu_0} = \frac{R(t_0)}{R(t_e)} \equiv 1 + z, \tag{9.9}$$

where we have defined the **cosmological redshift**, z. Thus, the further in the past that the light we receive was emitted (i.e., the more distant a source), the more the light is redshifted, in proportion to the ratio of the scale factors today and then. This, therefore, is the origin of Hubble's law.

Just like Doppler shift, the cosmological redshift of a distant object can be found easily by obtaining its spectrum and measuring the wavelengths of individual spectral features, either in absorption or in emission, relative to their laboratory wavelengths. Note, however, that cosmological redshift is distinct from Doppler, transverse-Doppler, and gravitational redshifts. The cosmological redshift of objects that are comoving with the Hubble flow is the result of the expansion of the scale of the Universe that takes place between emission and reception of a signal. In an expanding Universe (such as ours), $R(t_0) > R(t_e)$ always, and therefore z is always a redshift (rather than a blueshift). Indeed, it is found observationally that, beyond a distance of about 20 Mpc, all sources of light, without exception,

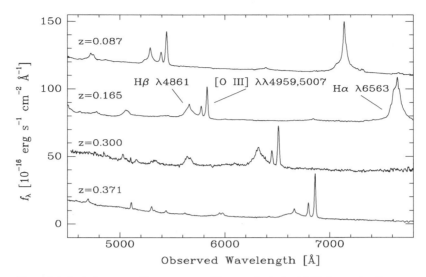

Figure 9.1 Optical spectra of four quasars, with cosmological redshifts increasing from top to bottom, as marked. Note the progression to the red of the main emission lines, which are indicated. The width of the Balmer lines is the result of Doppler blueshifts and redshifts about the line centers, due to internal motions of the emitting gas, under the influence of the central black holes powering the quasars. The [O III] lines are narrower because they are emitted by gas with smaller internal velocities. Data credit: S. Kaspi et al. 2000, *Astrophys. J.*, 533, 631.

are redshifted.[1] In addition to the cosmological redshift, the spectra of distant objects can be affected by (generally smaller) redshifts or blueshifts due to the other effects. Figure 9.1 shows the spectra of several distant quasars (objects that were discussed in section 6.3). Note the various redshifts by which the emission lines of each quasar (hydrogen Balmer Hα and Hβ, and the doublet [O III]$\lambda\lambda$ 4959, 5007 are the most prominent) have been shifted from their rest wavelengths by the cosmological expansion.

We have seen that the evolution of the scale factor, $R(t)$, depends on the parameters that describe the Universe, H_0, k, Ω_m, and Ω_Λ. This suggests that, if we could measure $R(t)$ at different times in the history of the Universe, we could deduce what kind of a universe we live in. In practice, it is impossible to measure $R(t)$ directly. However, the cosmological redshift z of an object gives the ratio between the scale factors today and at the time the light was emitted. We can therefore deduce the cosmological parameters by measuring properties of distant objects that depend on $R(t)$ through the redshift. Two such properties that have been particularly useful are the flux from an object and its angular size. Models with different cosmological parameters make different predictions as to how

---

[1] Nearby objects, such as Local Group galaxies and the stars in the Milky Way, are not receding with the Hubble flow (nor will they in the future) because they are bound to each other and to us. Similarly, the stars themselves, the Solar System, the Earth, and our bodies do not expand as the Universe grows. It is a common misconception that the "driving force" of the cosmological recession is the "expansion of space itself." In fact, galaxies are receding from us simply because they were doing so in the past, i.e., they have initial recession velocities and inertia (although now they are aided by dark energy—see below). A massless test particle placed at rest at any distance from us would not join the Hubble flow.

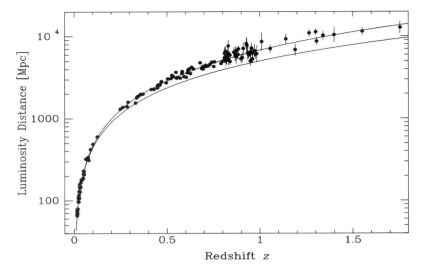

Figure 9.2 A Hubble diagram extending out to redshift $z \approx 1.7$, based on type Ia super-novae. Note that redshift now replaces velocity (compare to Fig. 7.6) and the *luminosity distances* to these standard candles are now plotted on the vertical axis. The top and bottom curves give the expected relations for cosmologies with $\Omega_m = 0.3$, $\Omega_\Lambda = 0.7$, and $\Omega_m = 1$, $\Omega_\Lambda = 0$, respectively. The data favor the top curve, indicating a cosmology currently dominated by dark energy. The calculation of the curves is outlined in Problems 4–7. Data credits: A. Riess et al. 2004, *Astrophys. J.*, 607, 665, and P. Astier et al., 2006, *Astron. Astrophys.*, 447, 31.

these observables change as a function of redshift. Measuring the flux from a "standard candle" to derive a "distance," and plotting the distance vs. the "velocity," is, of course, the whole idea behind the Hubble diagram. Now, however, we realize that cosmological redshift is distinct from Doppler velocity. Furthermore, in a curved and expanding space, "distance" can be defined in a number of different ways, and will depend on the properties and history of that space. Nevertheless, observables (e.g., the flux from an object of a given luminosity, or the angular size of an object of a given physical size, at some redshift) can be calculated straightforwardly from the FRW metric and the Friedmann equations and compared to the observations. We will work out examples of such calculations in section 9.3, and in Problems 4–7 at the end of this chapter.

In recent years, the Hubble diagram, based on type Ia supernovae serving as standard candles, has been measured out to beyond a redshift $z = 1$, corresponding to a time when the Universe was about half its present age. Figure 9.2 shows an example. The intrinsic luminosity of the supernovae at maximum light, compared to their observed flux, permits us to define a cosmological distance called **luminosity distance**,

$$D_L \equiv \left(\frac{L}{4\pi f} \right)^{1/2} . \tag{9.10}$$

The observed supernova fluxes (or, equivalently, their luminosity distances) vs. redshift are best reproduced by a model in which the Universe is currently in an accelerating stage, into which it transited (from the initial deceleration) at a time corresponding to about $z \sim 1$. If one assumes a flat, $k = 0$, Universe (for which the evidence will be presented in section 9.3), the data indicate $\Omega_m \approx 0.3$ and $\Omega_\Lambda \approx 0.7$. If this is true, the dynamics of the Universe are currently dominated by a "dark energy" of unknown source and nature that is causing the expansion to accelerate. The cosmological constant case, treated in section 8.5, is one possible form of the dark energy.

In the derivation of cosmological redshift, above, we considered the propagation of individual wavefronts of light. Instead, we could have discussed the propagation of, say, individual photons, or of brief light flashes, but would have gotten the same result: the time interval between emission of consecutive photons or light signals appears lengthened to the observer by a factor $1 + z$. Thus, in addition to cosmological redshift, light signals will undergo **cosmological time dilation**. For example, if a source at redshift z is emitting photons at a certain wavelength and at some rate, not only will an observer see the wavelength of every photon increased by $1 + z$, but the photon arrival rate will also be lower by $(1 + z)$. Both of these effects will reduce the observed energy flux, in addition to the reduction due to geometrical ($4\pi \times \text{distance}^2$) dilution (see Problem 3).

9.2 The Cosmic Microwave Background

Since the mean density of the Universe increases monotonically as one goes back in time,[2] there must have been an early time when the density was high enough such that the mean free path of photons was small, and baryonic matter and radiation were in thermodynamic equilibrium. The radiation field then had a Planck spectrum. Since the energy density of radiation changes with the scale factor as (Eq. 8.40)

$$\rho \propto R^{-4}, \tag{9.11}$$

but this energy density also relates to a temperature as

$$\rho = aT^4, \tag{9.12}$$

we can consider a **temperature** of the Universe at this stage, which varied as

$$T \propto \frac{1}{R}. \tag{9.13}$$

Therefore, early enough, the Universe was not only dense but also hot. At some stage, the temperature must have been high enough such that all atoms were constantly being ionized. The main source of opacity was then electron scattering. Going forward in time now, the temperature declined, and at $T \sim 3000$ K, few of the photons in the radiation field, even in its high-energy tail, had the energy required to ionize a hydrogen atom. Most of the

[2] In principle, models with a large enough positive cosmological constant permit a currently expanding Universe that had, in its past, a minimum R that is greater than zero, and thus no initial singularity. At times before the minimum, the Universe would have been contracting. In such a universe, as one looks to larger and larger distances, objects at first have increasing redshifts, as usual. However, beyond some distance, objects begin having progressively smaller redshifts, and eventually blueshifts. Such a behavior is contrary to observations.

electrons and protons then recombined. Once this happened, at a time $t_{rec} = 400,000$ yr after the Big Bang, the major source of opacity disappeared, and the Universe became transparent to radiation of most frequencies.[3] As we look to large distances in any direction in the sky, we look back in time, and therefore at some point our sightline must reach the **surface of last scattering**, beyond which the Universe is opaque.

The photons emerging from the last-scattering surface undergo negligible additional scattering and absorption until they reach us. Their number density therefore decreases, as the Universe expands, inversely with the volume, as R^{-3}. In addition, the energy of every photon is reduced by R^{-1} due to the cosmological redshift. The photon *energy* density therefore continues to decline as R^{-4}. Furthermore, the spectrum keeps its Planck shape, even though the photons are no longer in equilibrium with matter. To see this, consider that every photon gets redshifted from its emitted frequency ν to an observed frequency ν' according to the transformation

$$\nu' = \frac{\nu}{1+z}, \quad d\nu' = \frac{d\nu}{1+z}. \tag{9.14}$$

Next, recall the form of the Planck spectrum,

$$B_\nu = \frac{2h\nu^3}{c^2} \frac{d\nu}{e^{h\nu/kT} - 1}. \tag{9.15}$$

Dividing by the energy of a photon, $h\nu$, we obtain the *number* density of photons per unit frequency interval,

$$n_\nu = \frac{2\nu^2}{c^2} \frac{d\nu}{e^{h\nu/kT} - 1}. \tag{9.16}$$

Since the number of photons is conserved, their density decreases by a factor $(1+z)^3$, and the new distribution will be

$$n'_{\nu'} = \frac{n_\nu}{(1+z)^3} = \frac{2\nu^2}{c^2} \frac{d\nu}{e^{h\nu/kT} - 1} \frac{1}{(1+z)^3} = \frac{2\nu'^2}{c^2} \frac{d\nu'}{e^{h\nu'/kT'} - 1}, \tag{9.17}$$

where

$$T' \equiv \frac{T}{1+z}. \tag{9.18}$$

In other words, the spectrum keeps the Planck form, but with a temperature that is reduced, between the time of recombination and the present, according to

$$T_{cmb} = \frac{T_{rec}}{1+z_{rec}}, \tag{9.19}$$

where z_{rec} is the redshift at which recombination occurs. A prediction of Big Bang cosmology is therefore that space today should be filled with a thermal photon distribution arriving from all directions in the sky.

---

[3] The ubiquitous presence of hydrogen atoms in their ground state made the Universe, at this point, very opaque to ultraviolet radiation with wavelengths shortward of Lyman-α.

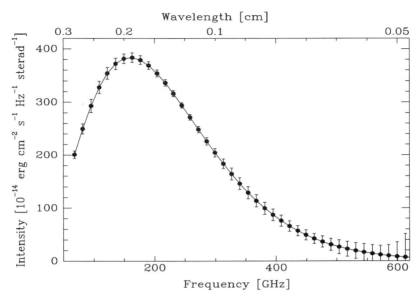

Figure 9.3 Observed spectrum of the cosmic microwave background, compared with a $T = 2.725$-K blackbody curve. The error bars shown are 500σ, so as to be discernible in the plot. Data credit: D. J. Fixsen et al. 1996, *Astrophys. J.*, 473, 576.

In the 1940s Gamow predicted, based on considerations of nucleosynthesis (which are discussed in the next section) that recombination must have occurred at $z_{rec} \sim 1000$, and hence the thermal spectrum should correspond to a temperature of a few to a few tens of degrees Kelvin (i.e., with a peak at a wavelength of order 1 mm, in the *microwave* region of the spectrum). This **cosmic microwave background** (CMB) radiation was discovered accidentally in 1965 by Penzias and Wilson, while studying sources of noise in microwave satellite communications. They translated the intensity they measured at a single frequency into a temperature, $T_{cmb} \approx 3$ K, by assuming that the radiation has a Planck spectrum and that the frequency is on the Rayleigh-Jeans side of the distribution[4] (Eq. 2.18), according to

$$B_\nu \approx \frac{2\nu^2}{c^2} kT. \tag{9.20}$$

Subsequent measurements, especially with several recent space-based experiments, have confirmed that the spectrum has a precise blackbody form, and have refined the temperature measurement to $T_{cmb} = 2.725 \pm 0.002$ K (see Fig. 9.3). Note that the CMB solves the Olbers paradox in a surprising way: every line of sight does indeed reach an ionized surface with a temperature similar to that of the photosphere of a star. Despite our being inside such an oven, we are not grilled because the expansion of the Universe dilutes the radiation emitted by this surface, and shifts it to harmless microwave energies.

---

[4] As opposed to the thermal flux from a star of unknown surface area, for which a temperature cannot be deduced from one or more measurements solely on the Rayleigh-Jeans side, the CMB is an **intensity**, i.e., an energy flux per unit solid angle on the sky, and it is completely specified for a blackbody of a given temperature. A temperature derived in this way is called by radio astronomers a **brightness temperature**.

The photon number density due to the CMB is

$$n_{\gamma,\mathrm{CMB}} \sim \frac{aT^4}{2.8kT} = \frac{7.6 \times 10^{-15} \text{ cgs} \times (2.7 \text{ K})^3}{2.8 \times 1.4 \times 10^{-16} \text{ erg K}^{-1}} = 400 \text{ cm}^{-3}. \tag{9.21}$$

Let us see that this is much larger than the cosmic mean number density of photons originating from stars. If n_{gal} is the mean number density of L_* galaxies, then at a typical point in the Universe the flux of starlight from galaxies within a spherical shell of thickness dr at a distance r from this point is

$$df = \frac{L_* n_{\mathrm{gal}} 4\pi r^2 dr}{4\pi r^2} = L_* n_{\mathrm{gal}} dr. \tag{9.22}$$

For a rough, order-of-magnitude, estimate of the total flux from galaxies at all distances, let us ignore the Universal expansion, possible curvature of space, and evolution with time of L_* and n_{gal}, and integrate from $r = 0$ to $r = ct_0$, where t_0 is the age of the Universe. Then the total flux is $f = L_* n_{\mathrm{gal}} ct_0$. Stars produce radiation mostly in the optical/IR range, with photon energies or order $h\nu_{\mathrm{opt}} \sim 1$ eV. The stellar photon density is about $1/c$ the photon flux. Thus,

$$\begin{aligned}
n_{\gamma,*} &\sim \frac{L_* n_{\mathrm{gal}} t_0}{h\nu_{\mathrm{opt}}} \approx \frac{10^{10} L_\odot \times 10^{-2} \text{ Mpc}^{-3} \times 14 \text{ Gyr}}{1 \text{ eV}} \\
&= \frac{10^{10} \times 3.8 \times 10^{33} \text{ erg s}^{-1} \times 10^{-2} \times (3.1 \times 10^{24} \text{ cm})^{-3} \times 4.4 \times 10^{17} \text{ s}}{1.6 \times 10^{-12} \text{ erg}} \\
&\approx 4 \times 10^{-3} \text{ cm}^{-3}.
\end{aligned} \tag{9.23}$$

Thus, there are of order 10^5 CMB photons for every stellar photon.[5]

The present-day baryon mass density is about 4% of the critical closure density, ρ_c. The mean baryon *number* density is therefore

$$n_B \approx \frac{0.04\rho_c}{m_p} \approx \frac{0.04 \times 9.2 \times 10^{-30} \text{ g cm}^{-3}}{1.7 \times 10^{-24} \text{ g}} = 2 \times 10^{-7} \text{ cm}^{-3}. \tag{9.24}$$

(Less than one-tenth of these baryons are in stars, and the rest are in a very tenuous intergalactic gas.) The baryon-to-photon ratio is therefore

$$\eta \equiv \frac{n_B}{n_\gamma} \approx 5 \times 10^{-10}. \tag{9.25}$$

Thus, although the *energy* density due to matter is much larger than that due to radiation (Eqs. 8.65 and 8.66), the *number* density of photons is much larger than the mean number density of baryons.

---

[5] The mean stellar photon density above is, of course, not representative of the stellar photon density on Earth, which is located inside an L_* galaxy, very close to an $L_\odot$ star. The daylight solar photon density on Earth (see Eq. 3.8) is 10^{10} times greater than the mean stellar value for the Universe, found above, and is thus also much greater than the CMB photon density.

9.3 Anisotropy of the Microwave Background

The temperature of the CMB, $T = 2.725$ K, is extremely uniform across the sky. There is a small **dipole** in the CMB sky, arising from the Doppler effect due mostly to the motion of the Local Group (at a velocity of ≈ 600 km s^{-1}) relative to the comoving cosmological frame. Apart from the dipole, the only deviations from uniformity in the CMB sky are temperature **anisotropies**, i.e., regions of various angular sizes with temperatures different from the mean, with fluctuations having root-mean-squared $\delta T = 29 \mu$K, or

$$\frac{\delta T}{T} \sim 10^{-5}. \tag{9.26}$$

Figure 9.4 shows a map of these temperature fluctuations. The extreme isotropy of the appearance of the Universe at $z \sim 1000$ is an overwhelming justification of the assumption of homogeneity and isotropy inherent to the cosmological principle. However, this extreme isotropy raises the questions of *why* and *how* the Universe can appear so isotropic. At the time of recombination, the horizon size—the size of a region in space across which light can propagate since the Big Bang (see chapter 8, Problems 1–3)—corresponded to a physical region that subtends only about 2° on the sky today. Thus, different regions separated by more than $\sim 2°$ could not have been in causal contact by t_{rec}, and therefore it is surprising that they would have the same temperature to within 10^{-5}. CMB photons from opposite directions on the sky have presumably never been in causal contact until now, yet they have almost exactly the same temperature.

The currently favored explanation for this "horizon problem" is that, very early during the evolution of the Universe, in the first small fraction of a second, there was an epoch of **inflation**. During that epoch, a vacuum energy density with negative pressure caused an exponential expansion of the scale factor, much like the second acceleration epoch that, apparently, we are in today. The inflationary expansion led causally connected regions to expand beyond the size of the horizon at that time. All the different parts of the microwave sky we see today were, in fact, part of a small, causally connected region before inflation. The cause and details of inflation are still a matter of debate, but most versions of the theory predict that, today, space is almost exactly flat (i.e., $\Omega_m + \Omega_\Lambda$ is very close to 1). We will see now that this prediction is strongly confirmed by the observed characteristics of the anisotropies.

The temperature anisotropies in the CMB arise through a number of processes, but at their root are small-amplitude inhomogeneities in the nearly uniform cosmic mass distribution. These inhomogeneities are set up at the end of the inflationary era, and their characteristics are yet another prediction of inflation theories. Most of the mass density at that time, as now, is in a nonbaryonic, pressureless, dark matter. Mixed with the dark matter, and sharing the same inhomegeneity pattern, is a relativistic gas of baryons and radiation. The photon–baryon gas therefore has an equation of state that is well described by

$$P = \tfrac{1}{3}\rho c^2. \tag{9.27}$$

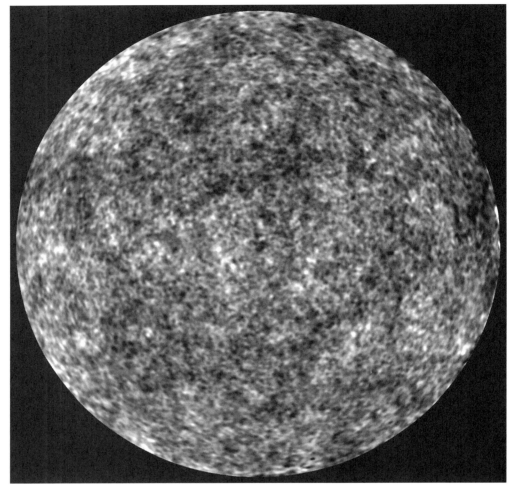

Figure 9.4 A half-sky (2π steradians) map of the temperature of the CMB sky. The typical relative fluctuations in the temperature, as coded by the gray scale (white is hot, black is cold), are of order 10^{-5}. Note the characteristic sizes of the hot and cold spots, $\approx 0.4°$. As described in the text, this size provides a "standard ruler" with which the geometry of space can be measured. Foreground microwave emission from the Milky Way has been subtracted from the image, as well as the CMB "dipole" anisotropy due to the motion of the Local Group relative to the comoving cosmological frame. Photo credit: NASA and the WMAP Science Team.

The speed of sound is then

$$c_s = \sqrt{\frac{dP}{d\rho}} = \frac{c}{\sqrt{3}}. \tag{9.28}$$

The mass density inhomogeneities have a spatial spectrum with power spread continuously among all Fourier components, i.e., they have no single physical scale. (The particular shape of the Fourier spectrum is, as noted above, a prediction of inflationary theories.) The gravitational potential of the inhomogeneities attracts the baryon–photon fluid, which is compressed in the denser regions and more tenuous in the underdense regions. However, the pressure of the fluid opposes the compression, and causes an expansion that stops only after the density has "overshot" the equilibrium density and the gas in the originally over-dense region has become underdense. Thus, periodic expansion and contraction of the various fluid regions ensues. This means that "standing" sound waves of all wavelengths represented in the spatial Fourier spectrum of the density inhomogeneities are formed in the photon–baryon gas.[6] Their periods τ and wavelengths λ are related by

$$\tau = \frac{\lambda}{c_s}. \tag{9.29}$$

When the Universe emerges from the inflationary era, at an age of a small fraction of a second, these **acoustic oscillations** are stationary and therefore they begin everywhere **in phase**. Consider now an overdense or underdense region. One of the Fourier modes that composes the region, and the fluid oscillations that it produces, has a wavelength that corresponds to a half-period of $t_{\rm rec}$,

$$\lambda = 2c_s t_{\rm rec} = \frac{2ct_{\rm rec}}{\sqrt{3}}, \tag{9.30}$$

where $t_{\rm rec}$ is the cosmic time when recombination occurs. At $t_{\rm rec}$, the baryon–photon fluid in this particular mode will have executed one-half of a full density oscillation, and will have just reached its maximal rarefaction or compression, where it will be colder or hotter, respectively, than the mean. At that time, however, the baryons and photons decouple, and the imprint of the cool (rarified) and hot (compressed) regions of the mode is frozen onto the CMB radiation field, and appears in the form of spots on the CMB sky with temperatures that are lower or higher than the mean. Similarly, higher modes that have had just enough time, between $t = 0$ and $t = t_{\rm rec}$, to undergo one full compression and one full rarefaction, or two compressions and a rarefaction, etc., will also be at their hottest or coldest at time $t_{\rm rec}$. The CMB sky is therefore expected to display spots having particular sizes. Stated differently, the fluctutation power spectrum of the CMB sky should have discrete peaks at these favored spatial scales.

   In reality, the picture is complicated by the fact that several processes, other than adia-batic compression, affect the gas temperature observed from each point. However, all these effects can be calculated accurately, and a prediction of the power spectrum can be made for a particular cosmological model. It turns out that measurement of the angular scales at the positions of the **acoustic peaks** in the power spectrum, and their relative heights, can determine most of the parameters describing a cosmological model. Let us see how this works for one example—the angular scale of the first acoustic peak as a measure of the global curvature of space.

   As seen in Eq. 9.30, the physical scale of the first acoustic peak is the *sound-crossing horizon* at the time of recombination. It therefore provides an excellent "standard ruler" at

---

[6] The waves that are formed are not, strictly speaking, standing waves, since they do not obey boundary conditions. They do resemble standing waves in the sense that a given Fourier component varies in phase at all locations. However, the superposition of all these waves is not a standing wave pattern, and does not have fixed nodes.

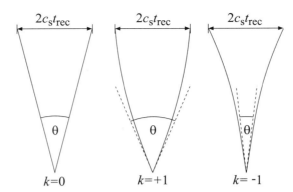

Figure 9.5 The angular diameter of the sound-crossing horizon (measurable from the size of the hot and cold spots in temperature anisotropy maps of the microwave sky), as it appears to observers in different space geometries. In a $k = 0$ universe ("flat" space), the spots subtend on the sky an angle θ given by Euclidean geometry. In a $k = +1$ Universe, the angles of a triangle with sides along geodesics sum to $>180°$. Since light follows a geodesic path, the converging light rays from the two sides of a CMB "spot" will bend, as shown, along their path, and θ will appear larger than in the $k = 0$ case. For negative space curvature, the angles in the triangle sum to $<180°$, and θ is smaller than in the flat case.

a known distance. The angle subtended on the sky by this standard ruler (i.e., the angle of the first peak) can be predicted for every geometry (i.e., curvature) of space. Comparison to the observed angle thus reveals directly what that geometry is (see Fig. 9.5).

Consider, for example, a flat ($k = 0$) cosmology with no cosmological constant. We wish to calculate the angular size on the sky, as it appears today, of a region of physical size (Eq. 9.30)

$$D_s = \frac{2ct_{\text{rec}}}{\sqrt{3}} = \frac{2 \times 400,000 \text{ ly}}{\sqrt{3}} = 140 \text{ kpc}, \tag{9.31}$$

from which light was emitted at time t_{rec}. Between recombination and the present time, the Universal expansion is matter-dominated, with $R \propto t^{2/3}$ for this model, i.e.,

$$\frac{R}{R_0} = \left(\frac{t}{t_0}\right)^{2/3} = \frac{1}{1+z}, \tag{9.32}$$

and hence we can also write D_s as

$$D_s = \frac{2ct_0}{\sqrt{3}}(1 + z_{\text{rec}})^{-3/2}. \tag{9.33}$$

The angle subtended by the region equals its size, divided by its distance to us *at the time of emission* (since that is when the angle between rays emanating from two sides of the region was set). As we are concerned with observed angles, the type of distance we are interested in is the distance that, when squared and multiplied by 4π, will give the area of the sphere centered on us and passing through the said region. If the comoving radial coordinate of the surface of last scattering is r, the required distance is *currently* just $r \times R_0$, and is called the **proper-motion distance**. (For $k = 0$, the proper distance and the proper motion distance are the same, as can be seen from Eq. 8.10.) The proper motion distance can again be found by solving for the null geodesic in the FRW metric (see Eq. 9.2),

$$\int_{t_{rec}}^{t_0} \frac{cdt}{R(t)} = \int_0^r \frac{dr}{\sqrt{1 - kr^2}}. \tag{9.34}$$

Setting $k = 0$, and substituting

$$R(t) = R_0 \left(\frac{t}{t_0} \right)^{2/3}, \tag{9.35}$$

we integrate and find

$$rR_0 = 3ct_0 \left[1 - \left(\frac{t_{rec}}{t_0} \right)^{1/3} \right] = 3ct_0[1 - (1 + z_{rec})^{-1/2}]. \tag{9.36}$$

However, at the time of emission, the scale factor of the Universe was $1 + z$ times smaller. The so-called **angular-diameter distance** to the last scattering surface is therefore

$$D_A = \frac{rR_0}{1 + z} = 3ct_0[(1 + z_{rec})^{-1} - (1 + z_{rec})^{-3/2}]. \tag{9.37}$$

The angular size of the sound-crossing horizon at the recombination era in a $k = 0$ cosmology is thus expected to be

$$\theta = \frac{D_s}{D_A} = \frac{2ct_0(1 + z_{rec})^{-3/2}}{3\sqrt{3}ct_0[(1 + z_{rec})^{-1} - (1 + z_{rec})^{-3/2}]}$$

$$= \frac{2}{3\sqrt{3}[(1 + z_{rec})^{1/2} - 1]}. \tag{9.38}$$

Since recombination occurs at $T_{rec} \approx 3000$ K, and the current CMB temperature is 2.7 K, $z_{rec} \approx 1100$, and

$$\theta \approx 0.012 \text{ radian} = 0.7°. \tag{9.39}$$

For this particular cosmological model ($k = 0$, $\Omega_\Lambda = 0$), this will be the angular scale of the first acoustic peak in the Fourier spectrum of the CMB fluctuations. The hot and cold "spots" in CMB sky maps will correspond to half a wavelength, i.e., will have half this angular size, or somewhat smaller than the diameter of the full Moon (half a degree). In a negatively curved geometry, where the angles of a triangle add up to less than 180°, the angle subtended by the standard ruler of length $2c_s t_{rec}$ will be smaller than in a flat geometry. In a positively curved Universe, this angle will appear larger than in the flat case.

Recent measurements of the CMB fluctuation power spectrum provide spectacular confirmation of the expected acoustic peaks (see Fig. 9.6). When compared to more

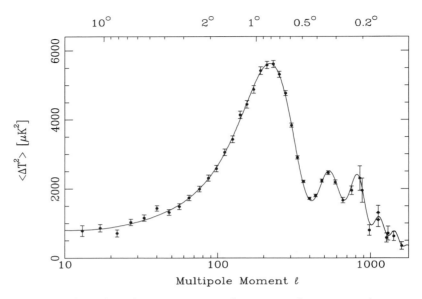

Figure 9.6 Observed angular power spectrum of temperature fluctuations in the CMB. The top axis shows the angular scales corresponding to the spherical harmonic multipoles on the bottom axis. The curve is based on a detailed calculation of the fluctuation spectrum using values for the various cosmological parameters that give the best fit to the data. Note the clear detection of acoustic peaks, with the first peak on a scale $\theta \approx 0.8°$, indicating a flat space geometry. Data credits: NASA/WMAP, CBI, and ACBAR collaborations.

sophisticated calculations that account for all the known effects that can influence the temperature anisotropies, the location of the first peak indicates a nearly flat space geometry, with

$$\Omega_m + \Omega_\Lambda = 1.02 \pm 0.02. \tag{9.40}$$

Note that a region with the diameter of the sound-crossing horizon has, between recombination and the present, expanded by $1 + z_{\rm rec} = 1100$, and hence encompasses today (i.e., has a *comoving diameter*) 140 kpc $\times$ 1100 = 150 Mpc. Thus, the CMB hot and cold spots correspond to regions that, today, are quite large.

Among a number of other cosmological parameters that are determined by analysis of the observed CMB anisotropy power spectrum are

$$\Omega_m \approx 0.3, \tag{9.41}$$

which together with Eq. 9.40 confirms the result found from the Hubble diagram of type Ia supernovae, that the dynamics of the Universe are currently dominated by a cosmological constant with

$$\Omega_\Lambda \approx 0.7. \tag{9.42}$$

If one assumes that the Universe is exactly flat, then the CMB results also give a precise age of the Universe

$$t_0 = 13.7 \pm 0.2 \text{ Gyr}, \tag{9.43}$$

and a density in baryons

$$\Omega_B = 0.044 \pm 0.004. \tag{9.44}$$

The mere existence of acoustic peaks in the power spectrum means that density perturbations existed long before the time of recombination, i.e., they were *primordial*, and that
they had wavelengths much longer than the horizon size at the time they were set up.
Inflation is the only theory that currently predicts, based on causal physics, the existence
of primordial, *superhorizon-size*, perturbations. The observation of the acoustic peaks can
therefore be considered as another successful prediction of inflation.

The large density inhomogeneities we see today—stars, galaxies, and clusters—formed
from the growth of the initial small fluctuations, the traces of which are observed in the
CMB. The gravitational pull of small density enhancements attracted additional mass, at
the expense of neighboring underdense regions. The growing clumps of dense matter
merged with other clumps to form larger clumps. This growth of structure by means of
gravitational instability operated at first only on the nonbaryonic dark-matter fluctuations,
but not the baryons, which were supported against gravitational collapse by radiation
pressure. Once the expansion of the Universe became matter-dominated, the dark-matter
density perturbations could begin to grow at a significant rate. Finally, after recombination,
the baryons became decoupled from the photons and their supporting radiation pressure,
and the perturbations in the baryon density field could also begin to grow. The details
and specific path according to which structure formation proceeds is still the subject of
active research. Nevertheless, it is clear that, once the first massive stars formed (ending
the period sometimes called the *Dark Ages*), they reionized most of the gas in the Universe. Based again on analysis of the CMB, current evidence is that this occurred during
some redshift in the range between ~ 6 and 20, when the Universe was 150–750 Myr
old.

By this time, the mean matter density was low enough that the newly liberated electrons
were a negligible source of opacity, and hence the Universe remained transparent (see
Problem 2). Direct evidence that most of the gas in the Universe is, at $z \sim 6$ and below,
almost completely ionized, comes from the fact that objects at those redshifts are visible
at UV wavelengths shorter than Lyman-α; even a tiny number of neutral hydrogen atoms
along the line of sight would suffice to completely absorb such UV radiation, due to the
very large cross section for absorption from the ground state of hydrogen (often called
resonant absorption). Most of the gas in the intergalactic medium (which is the main current
repository of baryons) remains in a low-density, hot, ionized phase. The density of this
gas is low enough that the recombination time is longer than the age of the Universe, and
hence the atoms will never recombine.

9.4 Nucleosynthesis of the Light Elements

Looking back in time to even earlier epochs than those discussed so far, the temperature of
the Universe must have been high enough that electrons, protons, positrons, and neutrons

were in thermodynamical equilibrium. Since the rest-mass energy difference between a neutron and a proton is

$$(m_n - m_p)c^2 = 1.3 \text{ MeV}, \tag{9.45}$$

at a time $t \ll 1$ s, when the temperature was $T \gg 1$ MeV (10^{10} K), the reactions

$$e^- + p + 0.8 \text{ MeV} \rightleftharpoons \nu_e + n \tag{9.46}$$

and

$$\bar{\nu}_e + p + 1.8 \text{ MeV} \rightleftharpoons e^+ + n \tag{9.47}$$

could easily proceed in both directions. The ratio between neutrons and protons as a function of temperature can be obtained from statistical mechanics considerations via the **Saha equation**. For the case at hand, it takes the form

$$\frac{N_n}{N_p} = \left(\frac{m_n}{m_p}\right)^{3/2} \exp\left[-\frac{(m_n - m_p)c^2}{kT}\right]. \tag{9.48}$$

When $T \gg 1$ MeV, the ratio is obviously very close to 1. As the temperature decreases, the ratio also decreases, and protons outnumber the heavier neutrons. This decrease in the ratio could continue indefinitely, but when $T < 0.8$ MeV, the mean time for reaction 9.46 becomes longer than the age of the Universe at that epoch, $t = 2$ s. The reaction time can be calculated from knowledge of the densities of the different particles, the temperature, and the cross section, as outlined for stellar nuclear reactions in Eqs. 3.123–3.127. The long reaction timescale means that the neutrons and protons, which are converted from one to the other via this reaction are no longer in thermodynamic equilibrium.[7] This time is called **neutron freezeout**, since neutrons can no longer be created. The neutron-to-proton ratio therefore "freezes" at a value of $\exp(-1.3/0.8) = 0.20$. In the following few minutes, most of the neutrons become integrated into helium nuclei. This occurs through the reactions

$$n + p \rightarrow d + \gamma \tag{9.49}$$
$$p + d \rightarrow {}^3\text{He} + \gamma \tag{9.50}$$
$$d + d \rightarrow {}^3\text{He} + n \tag{9.51}$$
$$n + {}^3\text{He} \rightarrow {}^4\text{He} + \gamma \tag{9.52}$$
$$d + {}^3\text{He} \rightarrow {}^4\text{He} + p. \tag{9.53}$$

Some of the neutrons undergo beta decay into a proton and an electron before making it into a helium nucleus (the mean lifetime of a free neutron is about 15 min), and a small fraction is integrated into other elements. Numerical computation of the results of all the parallel nuclear reactions that occur as the Universe expands, and as the density and the temperature decrease, shows that, in the end, the ratio between neutrons inside ^{4}He and protons is about 1/7. Thus, for every 2 neutrons there are 14 protons. Since every ^{4}He

---

[7] At about the same time, neutrinos also *decouple* (i.e., cease to be in thermal equilibrium with the rest of the matter and the radiation), and the cosmic neutrino background is formed; see Problem 9.

nucleus has 2 neutrons and 2 protons, there are 12 free protons for every ^{4}He nucleus, or the ratio of helium to hydrogen atoms is 1/12. The mass fraction of ^{4}He will then be

$$Y_4 = \frac{4N(^4\text{He})}{N(\text{H}) + 4N(^4\text{He})} = \frac{4\frac{1}{12}}{1 + 4\frac{1}{12}} = \frac{1}{4}. \tag{9.54}$$

A central prediction of Big Bang cosmology is therefore that a quarter of the mass in baryons was synthesized into helium in the first few minutes.

Measurements of helium abundance in many different astronomical settings (stars, H II regions, planetary nebulae) indeed reveal a helium mass abundance that is consistent with this prediction. This large amount of helium could not plausibly have been produced in stars. On the other hand, the fact that the helium abundance is nowhere observed to be lower than ≈ 0.25 is evidence for the unavoidability of primordial helium synthesis, at this level, among all baryons during the first few minutes.

Apart from ^{4}He, trace amounts of the following elements are produced during the first minutes: deuterium (10^{-5}), ^{3}He (10^{-5}), ^{7}Li (10^{-9}), ^{7}Be (10^{-9}), and almost nothing else. The precise abundances of these elements depend on the baryon density, n_B, at the time of nucleosynthesis. As we have seen (Eqs. 8.40, 9.13), the radiation energy density declines as R^{-4}, but the temperature appearing in the Planck spectrum also declines as $T \propto 1/R$, both before and after recombination. Since the energy of the photons scales with kT, the photon **number** density declines as R^{-3}. Because baryons are conserved, their density also declines as R^{-3} when the Universe expands, and therefore the baryon-to-photon ratio (Eq. 9.25), $\eta \approx 5 \times 10^{-10}$, does not change with time. Since we know the CMB photon density today, n_γ, measurements of the abundances of the light elements in astronomical systems that are believed to be pristine, i.e., that have undergone minimal additional processing in stars (which can also produce or destroy these elements) lead to an estimate of the baryon density today. In units of the critical closure density, ρ_c,

$$\Omega_B = \frac{n_B m_p}{\rho_c} = \frac{\eta \, n_\gamma \, m_p}{\rho_c}. \tag{9.55}$$

The baryon density based on these measurements is

$$0.01 < \Omega_B < 0.05. \tag{9.56}$$

As already mentioned, a completely independent estimate of Ω_B comes from analyzing the fluctuation spectrum of CMB anisotropies. The relative amplitudes of the acoustic peaks in the spectrum depend on the baryon density and hence constrain it to

$$\Omega_B = 0.044 \pm 0.004, \tag{9.57}$$

in excellent agreement with the value based on element abundances. Note that both of these measurements tell us that, even though the mass density of the Universe is a good fraction of the closure value ($\Omega_m \approx 0.3$), only a about a tenth of this mass is in baryons, while the rest must be in a dark matter component of unknown nature. Furthermore, less than 1/10 of the baryons are in stars inside galaxies. The bulk of the baryons are apparently in a tenuous, hot, and ionized intergalactic gas—the large reservoir of raw material out of which galaxies formed. A small fraction of this gas is neutral, and can be observed by the absorption it produces in the spectra of distant quasars. This is discussed briefly in section 9.5.

Table 9.1 History and Parameters of the Universe

Curvature: $\Omega_m + \Omega_\Lambda = 1.02 \pm 0.02$

Mass density: $\Omega_{m,0} \approx 0.3$, consisting of

$\Omega_{B,0} = 0.044 \pm 0.004$ in baryons, and

$\Omega_{DM,0} \approx 0.25$ in dark matter

Dark energy: $\Omega_\Lambda \approx 0.7$

Time	Redshift z	Temperature $T(K)$	Event
$\sim 10^{-34}$ s	$\sim 10^{27}$	$\sim 10^{27}$	Inflation ends, $\Omega_m + \Omega_\Lambda \to 1$, causally connected regions have expanded exponentially, initial fluctuation spectrum determined.
2 s	4×10^9	10^{10}	Neutron freezeout, no more neutrons formed.
3 min	4×10^8	10^9	Primordial nucleosynthesis over—light element abundances set.
65,000 yr	3500	10^4	Radiation domination $\to$ mass domination, $R \sim t^{1/2} \to R \sim t^{2/3}$, dark-matter structures start growing at a significant rate.
400,000 yr	1100	3000	Hydrogen atoms recombine, matter and radiation decouple, Universe becomes transparent to radiation of wavelengths longer than Lyα, CMB fluctuation pattern frozen in space, baryon perturbations start growing.
$\sim 10^8$–10^9 yr	~ 6–20	~ 20–60	First stars form and reionize the Universe, ending the Dark Ages. The Universe becomes transparent also to radiation with wavelengths shorter than Lyα.
~ 6 Gyr	~ 1	~ 5	Transition from deceleration to acceleration under the influence of dark energy.
14 Gyr	0	2.725 ± 0.002	Today.

Table 9.1 summarizes the current view of the cosmological parameters and the history of the Universe.

9.5 Quasars and Other Distant Sources as Cosmological Probes

Quasars, which we discussed in section 6.3, are supermassive black holes accreting at rates that produce near-Eddington luminosities of 10^1–$10^4 L_*$. Their large luminosities make

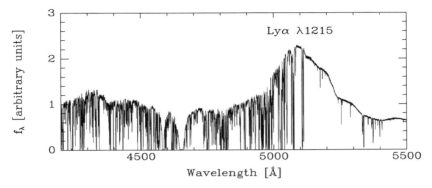

Figure 9.7 A high-resolution spectrum of a quasar at redshift $z = 3.18$, with the Lyman α emission line redshifted to 5080 Å. Note the *Lyman-α forest* of absorption lines starting from the peak of the emission line, and continuing in the blue (left) direction. These lines are due to Lyman-α absorption by neutral hydrogen atoms in gas clouds that are along the line of sight to the quasar, and hence at lower redshifts than the quasar. The few absorption lines to the red of the Lyman-α emission line peak are due to heavier elements and are associated with the system that produces the strong *damped Lyman-α* absorption observed at $\approx$4650 Å. Data credit: W. Sargent and L. Lu, based on observations with the HIRES spectrograph at the W. M. Keck Observatory.

quasars easily visible to large cosmological distances, and allow probing the assembly and accretion history of the central black holes of galaxies. As noted in chapter 6, luminous quasars are rare objects at present, and apparently most central black holes in nearby galaxies are accreting at low or moderate rates, compared to the rates that would produce a luminosity of L_E. However, quasars were much more common in the past, and their comoving space density reached a peak at an epoch corresponding to redshift $z \sim 2$ (i.e., about 10 Gyr ago). There is likely a connection between the growth and development of galaxies and of their central black holes, and quasar evolution may hold clues to deciphering this connection (see Problem 11). The most distant quasars currently known are at redshifts beyond $z = 6$, and are therefore observed less than 1 Gyr after the Big Bang. Models of structure formation suggest that the first galaxies began to assemble at about that time.

Since quasars are so luminous, they are also useful cosmological tools, in that they can serve as bright and distant sources of light for studying the contents of the Universe between the quasars and us. One such application is the study of **quasar absorption lines**. The light from all distant quasars is seen to be partially absorbed by numerous clouds of gas along the line of sight. A small fraction ($\sim 10^{-4}$) of the hydrogen in these clouds is neutral, and is manifest as a "forest" of redshifted absorption lines (mostly Lyman-α) in the spectrum of each quasar (see Fig. 9.7). Each absorption line is at the wavelength of Lyman-α redshifted according to the distance of the particular absorbing cloud. The absorption lines are therefore distributed in wavelength between the rest wavelength of Lyα at 1216 Å and the observed, redshifted Lyα wavelength of the quasar (say, $(1 + z)1216$ Å $= 3648$ Å, for a $z = 2$ quasar).

Apart from the hydrogen Lyman-α lines, additional absorption lines are detected. Absorption lines produced by heavier elements in the same clouds allow estimating the "metallicities" of these clouds, and reveal very low element abundances, i.e., the gas in the clouds has undergone little enrichment by stellar processes. It is in such clouds that the abundance of primordial deuterium can be measured and compared to Big Bang nucleosynthesis predictions (see section 9.4). The Lyman-α clouds are one component (a relatively cool one, with $T \sim 10^4$ K) of the **intergalactic medium**. Most of the intergalactic gas, however, is apparently in a hotter $T \sim 10^{5-6}$ K, more tenuous, component. Estimates of the total mass density of intergalactic gas find that the bulk of the baryons in the Universe is contained in this hot component, while less than about 10% of the baryons are in galaxies in the form of stars and cold gas.

Another application in which quasars serve as distant light sources for probing the intervening matter distribution is in cases where galaxies or galaxy clusters gravitationally lens quasars that are projected behind them, splitting them into multiple images.[8] Since the lensing objects in such cases are at cosmological distances ($\sim$1 Gpc), and the lensing masses are of order $10^{11} M_\odot$, the Einstein angle (Eq. 6.25), which gives the characteristic angular scale of the split images, is of order 1 arcsecond, i.e., resolved by telescopes at most wavelength bands, from radio through X-rays (see Fig. 9.8). Modeling of individual systems can reveal the shapes and forms of the mass distributions, both the dark and the luminous. The statistics of lensed quasars (e.g., measurement of the fraction of quasars that are multiply imaged by intervening galaxies) can provide information on the properties of the galaxy population and its evolution with cosmic time (see chapter 6, Problem 6). Not only quasars serve as background light sources for galaxy lenses—there are many known cases of galaxies that lens other galaxies that lie behind them (also shown in Fig. 9.8), and such systems can be used for the same applications.

In known systems in which a galaxy or a galaxy cluster operate as a powerful gravitational lens, one can turn the problem around and use the lens as a "natural telescope." Once the properties of the lens have been derived, based on the positions and relative magnifications of the lensed images of the bright background quasar or galaxy, one can search other regions of the lens that are then expected to produce high magnification for lensed images of additional background objects. This method of "searching under the magnifying glass" has been used to find and study galaxies with luminosities as low as $0.01 L_*$ out to redshifts $z \sim 6$, aided by the natural magnification of galaxy clusters.

With these and other techniques, it is hoped that a detailed and consistent picture of cosmic history will eventually emerge. Such an understanding would include the nature of dark matter and dark energy, their interplay with baryons and with supermassive black holes in the formation of the first stars and galaxies, the element enrichment of the interstellar and the intergalactic medium by generations of evolved stars and supernovae, and the evolution of galaxies and their constituents, all the way to the world as we see it today.

---

[8] Since galaxy mass distributions are generally not spherically symmetric, when they act as gravitational lenses they can split background sources into multiple images, rather than just deforming the sources into rings or splitting them into double images, as is the case for point masses and spherically symmetric masses.

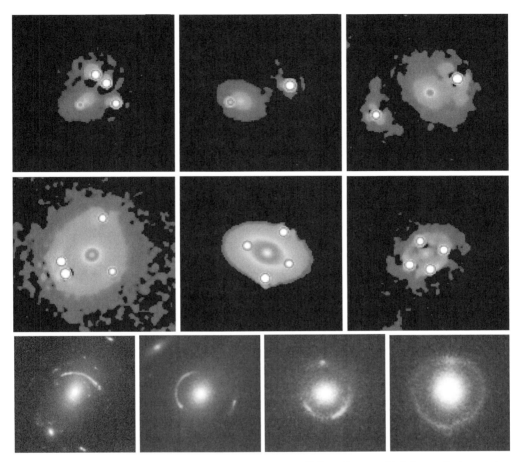

Figure 9.8 *Top two rows:* Examples of quasars that are gravitationally lensed into multiple images by intervening galaxies. In each case, the lens galaxy, at a redshift of $z \approx 0.04$–0.7, is the extended central object, and the two or four sources straddling it are the multiply lensed images of a background quasar, at $z \approx 1.7$–3.6. Panels are 5 arcseconds on a side. Some image processing has been applied, to permit seeing clearly both the bright, point-like, quasar images and the faint, extended lens galaxies. *Bottom row:* Examples of foreground galaxies that lens background galaxies into partial or full Einstein rings. In the cases shown, the foreground galaxies are at $z \approx 0.2$–0.4 and the background galaxies are at $z \approx 0.5$–1. Photo credits: The CASTLES gravitational lens database, C. Kochanek et al.; NASA, ESA, J. Blakeslee and H. Ford,; and NASA, ESA, A. Bolton, S. Burles, L. Koopmans, T. Treu, and L. Moustakas.

Problems

1. In an accelerating or decelerating Universe, the redshift z of a particular source will slowly change over time t_0, as measured by an observer.

   a. Show that the rate of change is

   $$\frac{dz}{dt_0} = H_0(1 + z) - H(z),$$

   where $H(z) \equiv \dot{R}_e/R_e$ is the Hubble parameter at the time of emission.

Hint: Differentiate the definition of redshift, $1 + z \equiv R_0/R_e$, with respect to t_0. Use the chain rule to deal with expressions such as dR_e/dt_0.

b. Show that, for a $k = 0$ universe with no cosmological constant, $H(z) = H_0(1 + z)^{3/2}$. For this model, and assuming $H_0 = 70 \text{ km s}^{-1}\text{Mpc}^{-1}$, evaluate the change in redshift over 10 years, for a source at $z = 1$, and the corresponding change in "recession velocity".

Answers: $\Delta z = -5.9 \times 10^{-10}$, $\Delta v = -18 \text{ cm s}^{-1}$.

2. At a redshift $z = 1100$, atoms were formed, the opacity of the Universe to radiation via electron scattering disappeared, and the cosmic microwave background was formed. Imagine a world in which atoms cannot form. Even though such a universe, by definition, will remain ionized forever, after enough time the density will decline sufficiently to make the universe transparent nonetheless. Find the redshift at which this would have happened, for a $k = 0$ universe with no cosmological constant. Assume an all-hydrogen composition, $\Omega_B = 0.04$, and $H_0 = 70 \text{ km s}^{-1}\text{Mpc}^{-1}$. Note that this calculation is not so farfetched. Following recombination to atoms at $z = 1100$, most of the gas in the Universe was reionized sometime between $z = 6$ and $z = 20$ (probably by the first massive stars that formed), and has remained ionized to this day. Despite this fact, the opacity due to electron scattering is very low, and our view is virtually unhindered out to high redshifts.

Hint: A "Universe transparent to electron scattering" can be defined in several ways. One definition is to require that the rate at which a photon is scattered by electrons, $n_e \sigma_T c$, is lower than the expansion rate of the Universe at that time, H (or, in other words, the time between two scatters is longer than the age of the Universe at that time). To follow this path (which is called *decoupling* between the photons and the hypothetical free electrons), express the electron density n_e at redshift z, by starting with the current baryon number density, $\Omega_B \rho_{cr,0}/m_p$, expressing $\rho_{cr,0}$ by means of H_0, and increasing the density in the past as $(1 + z)^3$. Similarly, write H in terms of H_0 and $(1 + z)$ (recall that $1 + z = R_0/R$, and in this cosmology, $R \propto t^{2/3}$ and $H \propto t^{-1}$). Show that decoupling would have occurred at

$$1 + z = \left(\frac{8\pi G m_p}{3\Omega_B H_0 \sigma_T c} \right)^{2/3},$$

and calculate the value of this redshift. Alternatively, we can find the redshift of the "last scattering surface" from which a typical photon would have reached us without further scatters. The number of scatters on electrons that a photon undergoes as it travels from redshift z to redshift zero is

$$\int_0^{l(z)} n_e(z)\sigma_T dl.$$

Express n_e, as above, in terms of Ω_B, H_0, and $1 + z$, replace dl with $c(dt/dz)dz$, using again $R \propto t^{2/3}$ to write dt/dz in terms of H_0 and $1 + z$. Equate the integral to 1, perform the integration, show that the last scattering redshift would be

$$1 + z = \left(\frac{4\pi G m_p}{\Omega_B H_0 \sigma_T c} \right)^{2/3},$$

and evaluate it.

Answers: $z = 65$; $z = 85$.

3. Show that the angular-diameter distance for a flat space ($k = 0$; Eq. 9.37) out to redshift z,

$$D_A = 3ct_0[(1 + z)^{-1} - (1 + z)^{-3/2}],$$

has a maximum with respect to redshift z, and find that redshift. The angular size on the sky of an object with physical size d is $\theta = d/D_A$. What is the implication of the maximum of D_A for the appearance of objects at redshifts beyond the one you found? Note that this peculiar behavior is simply the result of light travel time out to different distances in an expanding universe; an object at high redshift may have been closer to us at the time of emission than an object of the same size at a lower redshift, despite the fact that the high-redshift object is currently more distant.

4. a. Consider the energy flux of photons from a source with bolometric luminosity L and with proper-motion distance rR_0. The photons will be spread over an area $4\pi (rR_0)^2$. Explain why the observed energy flux will be

$$f = \frac{L}{4\pi (rR_0)^2 (1 + z)^2}.$$

   Hint: Consider the effects of redshift on the photon energy and cosmological time dilation on the photon arrival rate. This relation is used to define the *luminosity distance,* $D_L = rR_0(1 + z)$.
   b. Find $D_L(z)$ for a $k = 0$ universe without a cosmological constant. Plot, for this world model, the Hubble diagram, i.e., the flux vs. z, from an object of constant luminosity.

5. Show that in a Euclidean, nonexpanding, universe, the surface brightness of an object, i.e., its observed flux per unit solid angle (e.g., per arcsecond squared), does not change with distance. Then, show that in an expanding FRW universe, the ratio between the luminosity distance (see Problem 4) and the angular-diameter distance to an object is always $(1 + z)^2$. Use this to prove that, in the latter universe, surface brightness dims with increasing redshift as $(1 + z)^{-4}$. This effect makes extended objects, such as galaxies, increasingly difficult to detect at high z.

6. An object at proper-motion distance rR_0 splits into two halves. Each piece moves relative to the other, perpendicular to our line of sight, at a constant, nonrelativistic, velocity v. What is the the angular rate of separation, or "proper motion" between the two objects (i.e., the change of angle per unit time)?
   Hint: Recall that we are measuring an angle, and so require the angular-diameter distance, but we are also measuring a rate, which is affected by cosmological time dilation. You can now see why rR_0 is called the proper-motion distance.

7. Use the first Friedmann equation with a nonzero cosmological constant (Eq. 8.95) to show that, in a flat, matter-dominated Universe, the proper-motion distance is

$$rR_0 = \int \frac{cdz}{H_0\sqrt{\Omega_{m,0}(1+z)^3 + \Omega_{\Lambda,0}}}.$$

Use a computer to evaluate this integral numerically with $\Omega_{m,0} = 0.3$ and $\Omega_{\Lambda,0} = 0.7$, for values of z between 0 and 2. Plot the Hubble diagram, i.e., flux vs. z, from an object of constant luminosity, in this case, and compare to the curve describing $k = 0$, $\Omega_m = 1$ (Problem 4). You can now see how the Hubble diagram of type Ia supernovae can distinguish among cosmological models.

Hint: Set $k = 0$ in Eq. 8.95, replace ρ by $\rho_0 R_0^3/R^3$ (matter domination), divide both sides by H_0^2, and substitute the dimensionless parameters $\Omega_{m,0}$ and $\Omega_{\Lambda,0}$. Change variables from R to z with the transformation $1 + z = R_0/R$, and separate the variables z and t. Finally, use the FRW metric for $k = 0$: $cdt = Rdr = R_0/(1+z)dr$, and hence $rR_0 = \int (1+z)cdt$, to obtain the desired result.

8. Emission lines of hydrogen Hβ ($n = 4 \to 2$, $\lambda_{rest} = 4861$ Å) are observed in the spectrum of a spiral galaxy at redshift $z = 0.9$. The galaxy disk is inclined by $45°$ to the line of sight.

   a. The Hβ wavelength of lines from one side of the galaxy are shifted to the blue by 5 Å relative to the emission line from the center of the galaxy, and to the red by 5 Å on the other side. What is the galaxy's rotation speed?

   b. Analysis of the emission from the active nucleus of the galaxy reveals a total redshift of $z = 1$. If the additional redshift is gravitational, the result of the proximity of the emitting material to a black hole, find this proximity, in Schwarzschild radii.

   Hint: Note that all redshift and blueshift effects are multiplicative, e.g., $(1 + z_{total}) = (1 + z_{cosmological})(1 + v \sin i/c)$, or $(1 + z_{total}) = (1 + z_{cosmological})(1 + z_{gravitational})$.

   c. Find the age of the Universe at $z = 0.9$, assuming an expansion factor $R \propto t^{2/3}$, and a current age $t_0 = 14$ Gyr. What is the "lookback time" to the galaxy?

   Answers: 230 km s^{-1}; $11r_s$; lookback time 8.5 Gyr.

9. At some point back in cosmic time, the Universe was dense enough to be opaque to neutrinos. Then, as the Universe expanded, the density decreased until neutrinos could stream freely. A *cosmic neutrino background* (which is undetected to date) must have formed when this decoupling between neutrinos and normal matter occurred, in analogy to the CMB that results from the electron–photon decoupling at the time of hydrogen recombination. Find the temperature at which neutrino decoupling occurred. Assume in your calculation that decoupling occurs during the radiation-dominated era, photons pose the main targets for the neutrinos, neutrino interactions have an energy-dependent cross section

$$\sigma_{\nu\gamma} = 10^{-43} \text{ cm}^2 \left(\frac{E_\nu}{1 \text{ MeV}}\right)^2,$$

and the neutrinos are relativistic. Use a $k = 0$, $\Omega_\Lambda = 0$ cosmology.

Hint: Proceed by the first method of Problem 2, i.e., by requiring $n\sigma v = H$. Represent the "target" density, n, by aT^4/kT, where a is the Stefan-Boltzmann (or "radiation") constant. Use $\sigma_{\nu\gamma}$ for the cross section σ, but approximating E_ν as kT. The velocity v equals c, because the neutrinos and the target particles are relativistic. To represent H, use the first Friedmann equation,

$$H^2 = \frac{8\pi G \rho_{\rm rad}}{3c^2},$$

with $\rho_{\rm rad} = aT^4$.

Answer: $kT = 1$ MeV.

10. It has been found recently that every galactic bulge harbors a central black hole with a mass ~ 0.001 of the bulge mass. The mean space density of bulges having $10^{10} M_\odot$ is about 10^{-2} Mpc^{-3}.

   a. Find the mean density of mass in black holes, in units of $M_\odot$ Mpc^{-3}.

   b. If all these black holes were shining at their Eddington luminosities, what would be the luminosity density, in units of $L_\odot$ Mpc^{-3}? How does this compare to the luminosity density from stars?

   c. The observed luminosity density of quasars and active galaxies, averaged over cosmic time, is actually 100 times less than calculated in (b). If all central black holes have gone through an active phase, what does this imply for the total length of time that a black hole is "active"?

11. The most distant quasars currently known are at redshift $z \sim 6$, and have luminosities $L \sim 10^{47}$ erg s^{-1}.

   a. Find a lower limit to the mass of the black hole powering such a quasar, by assuming it is radiating at the Eddington limit.

   b. Find the age of the Universe at $z = 6$, assuming an expansion $R \propto t^{2/3}$ and a current age $t_0 = 14$ Gyr.

   c. Equate the Eddington luminosity $L_E(M)$ as a function of mass M to the luminosity of an accretion disk around a black hole with a mass-to-energy conversion efficiency of 0.06. This will give you a simple differential equation for $M(t)$, describing the growth of a black hole. Solve the equation (be careful with units).

   d. Suppose a black hole begins with a "seed" mass of $10 M_\odot$ and shines at the Eddington luminosity continuously. How long will it take the black hole to reach the mass found in (a)? By comparing to the result of (b), what is the minimum redshift at which accretion must begin?

   Answers: $\sim 10^9 M_\odot$; $t(z = 6) = 740$ Myr; $M = M_{\rm seed} \exp(t/\tau)$, with $\tau = 26$ Myr; 480 Myr, $z(t = 260$ Myr$) = 13$.

| **Appendix** | Recommended Reading and Websites |

Textbooks

The following textbooks are at the advanced-undergraduate or graduate level, and are a good place to pick up where this book takes off on the various topics.

Observational Techniques

Kitchin, C. R. (2003) *Astrophysical techniques* (4th ed.), Institute of Physics
McLean, I. S (1997) *Electronic imaging in astronomy*, Wiley

Radiative Processes

Rybicki, G. B., & Lightman, A. P. (1979) *Radiative processes in astrophysics*, Wiley-Interscience

Stellar Structure and Evolution

Prialnik, D. (2000) *An introduction to the theory of stellar structure and evolution*, Cambridge University Press
Phillips, A. C. (1994) *The physics of stars*, Wiley
Clayton, D. D. (1983) *Principles of stellar evolution and nucleosynthesis*, University of Chicago Press
Hansen, C. J., Kawaler, S. D., & Trimble, V. (2004) *Stellar interiors* (2nd ed.), Springer-Verlag.
Bahcall, J. N. (1989) *Neutrino astrophysics*, Cambridge University Press

Stellar Remnants and Accretion Physics

Shapiro, S. L., & Teukolsky, S. A. (1983) *Black holes, white dwarfs, and neutron stars: The physics of compact objects*, Wiley-Interscience

Interstellar Medium

Spitzer, L. (1978) *Physical processes in the interstellar medium*, Wiley-Interscience
Osterbrock, D. E., & Ferland, G. J. (2006) *Astrophysics of gaseous nebulae and active galactic nuclei* (2nd ed.), University Science Books

Galaxies

Binney, J. & Merrifield, M. (1998) *Galactic astronomy*, Princeton University Press
Binney, J. & Tremaine, S. (1987) *Galactic dynamics*, Princeton University Press

Active Galactic Nuclei

Peterson, B. M. (1997) *An introduction to active galactic nuclei*, Cambridge University Press
Krolik, J. H. (1999) *Active galactic nuclei*, Princeton University Press

General Relativity and Cosmology

There is a large assortment of good cosmology texts at many different levels. The following is a very partial list, more or less by increasing level.
Roos, M. (1997) *Introduction to cosmology*, Wiley
Liddle, A. (2003) *An introduction to modern cosmology*, Wiley
Ryden, B. (2003) *Introduction to cosmology*, Addison Wesley
Misner, C. W., Thorne, K. S., & Wheeler, J. A. (1973) *Gravitation*, Freeman
Peacock, J. A. (1999) *Cosmological physics*, Cambridge University Press
Weinberg, S. (1972) *Gravitation and cosmology*, Wiley
Peebles, P.J.E. (1993) *Principles of physical cosmology*, Princeton University Press
Kolb, E. W., & Turner, M. S. (1990) *The early universe*, Addison Wesley

Astronomical Data and Reference

A useful compendium of astronomical data and formulae is
Cox, A. N., editor (2000) *Allen's astrophysical quantities* (4th ed.), Springer-Verlag

Websites

The following useful internet websites are fairly well-established, and their addresses should therefore remain accurate for some time.

Pretty Pictures (and More)

The Hubble Space Telescope has obtained visually stunning images of many of the types of astronomical objects discussed in this book. See them at:
http://hubblesite.org

Equally beautiful pictures based on data from ground-based telescopes are displayed on the websites of the European Southern Observatory, the National Optical Astronomy Observatory, and the National Radio Astronomy Observatory:
http://www.eso.org/outreach/gallery/astro/
http://www.noao.edu/image_gallery/
http://www.nrao.edu/imagegallery

A refreshing way to start (or end) every day is to check out NASA's "Astronomy Picture of the Day":
http://antwrp.gsfc.nasa.gov/apod/

Research Papers

Almost all research papers in astrophysics written nowadays are posted by their authors on the arXiv e-print archive, from where they can be freely downloaded. The main address below has many mirror sites.

`http://xxx.arxiv.org/archive/astro-ph`

The published versions of all papers that have appeared in the major astrophysics journals are available online through the NASA Astrophysics Data System. Downloading the full versions of recently published (past few years) articles often requires a journal subscription. If your university library does not have a subscription, the final or near-final versions of the papers can usually also be found in the arXiv website above.

`http://adsabs.harvard.edu/abstract_service.html`

Astronomical Databases

Two useful sites for obtaining information on particular astronomical objects (positions, redshifts, photometry, literature, etc.) are the SIMBAD Astronomical Database and the NASA/IPAC Extragalactic Database (NED):

`http://simbad.u-strasbg.fr/Simbad`
`http://nedwww.ipac.caltech.edu`

Additional Materials

Corrections and updates to this book, downloadable versions of the figures (with halftones in their original color renditions), and instructions for teachers wishing to obtain a solutions manual for the problems, are available at:

`http://press.princeton.edu/titles/8457.html`

Index